Praise for *American Eclipse*

"A suspenseful narrative history about the last total solar eclipse to cross North America in the summer of 1878. . . . [David] Baron makes those three minutes seem transcendent."

—Maureen Corrigan, NPR's *Fresh Air*

"Baron, an award-winning journalist, uses exhaustive research to reconstruct a remarkable chapter of U.S. history. He tells the surprising story of how the eclipse spurred three icons of the 19th century—inventor Thomas Edison, planet hunter James Craig Watson, and astronomer and women's-rights crusader Maria Mitchell—to trek into the wild Western frontier to observe it."

—Lee Billings, *Scientific American*

"The stories of these three enterprising scientists reflect the ambition and intellectual curiosity of the United States in the late-nineteenth-century, when the country was trying to cement its place in the international scientific community."

—Concepción de León, *New York Times*

"Baron is an umbraphile—an eclipse enthusiast—and it shows in his vivid account. . . . [He] captures the celestial drama nicely, and there is human drama here, as well. . . . The book's achievement lies in taking the measure of what this conjunction of celestial bodies and historical figures reveals about America, 'as a society, a nation, a civilization,' circa July 1878."

—Evan Hepler-Smith, Wall Street Journal

"*American Eclipse* is an incredibly well written work of non-fiction. It is clearly the result of considerable research and careful thought. And it tells a great story."

—Simon Perks, BBC Sky at Night, Book of the Month

"*American Eclipse* is a social story dressed in scientific garb, more *This American Life* than *Scientific American*. . . . Baron breaks down the science with a patience learned from three decades in public radio."

—Graham Ambrose, Denver Post

"Baron spins a historical tale of the scientific personalities of the era, focused on three researchers who traveled to Colorado and Wyoming to witness totality for very different reasons. . . . Baron grounds the stories of these three in the context of the rise of American science, providing a deeply researched and original contribution."

—Alexandra Witze, Dallas Morning News

"An entertaining and highly readable account of the total eclipse of July 29, 1878, and three particularly notable scientific expeditions that set out to see and learn from it."

—Sarah Bryan Miller, St. Louis Post-Dispatch

"In his riveting account, Baron shows his ability to wade through slews of historical documents . . . to bring the [story] to life."

—Sheena McFarland, Salt Lake Tribune

"Baron's account is more historical than scientific, but the science that is described—e.g., spectroscopy, penumbras, and latitude and longitude measurements—is well rendered. . . . [M]oves at a fast clip and covers a lot of ground."   —Jennifer Carson, *Science*

"Baron tells a lively tale that places eclipse science in the historical context of the Wild West. . . . There's the brutality of a frontier lynching, the annoyance of lost railway luggage and the beauty of the astronomical event itself."   —Stuart Clark, *New Scientist*

"Baron's book does a wonderful job weaving together these accounts of early American ingenuity as the scientists headed west, determined to prove their prowess during two minutes of darkness."
         —Heather Goss, *Air & Space Magazine*

"In this delightfully readable work of science history, we see an ardent young republic testing its intellectual prowess on the world stage. Baron has chosen just the right moment, and peopled it with just the right characters. This fascinating portrait of the Gilded Age is suffused with the peculiar magic and sense of awe that have always attended eclipses, those fraught few minutes when day becomes night, time stands still—and anything seems possible."
         —Hampton Sides, *New York Times* best-selling author of
Blood and Thunder

"David Baron contracted an incurable case of umbraphilia twenty years ago in Aruba. Fortunately for readers, Baron's fever stokes his account of the first great American eclipse, in 1878, while priming us for the next one—and the next, and the next."
         —Dava Sobel, author of *The Glass Universe*

"David Baron beautifully captures the awe, the magic, and the mystery of one particular eclipse, an event in 1878 that spurred on America to embrace the sciences. A superb contribution to the history of astronomy."
         —Marcia Bartusiak, author of *Einstein's Unfinished Symphony*

"A wonderful book, bringing lessons from the past to the present. In exceptionally clear and interesting prose, Baron brings nineteenth-century personalities to life, showing how men and, unusually, a female astronomy professor of that time observed the total solar eclipse of 1878."

—Jay Pasachoff, Field Memorial Professor of Astronomy at Williams College

"Lucidly melds science, ambition, policy, technology, the interplay of personality and practice, and the immediacy of experience. The book is marked by wonderful, eye-opening surprises, notably Edison's enthusiasm for and participation in the observation of the eclipse and the independent expedition of Maria Mitchell and her crew in the face of their exclusion from the effort."

—Daniel Kevles, author of *The Physicists*

"Brilliantly researched and beautifully crafted, *American Eclipse* conveys historical discoveries and scientific obsessions with the verve and excitement of a work of fiction. David Baron's vivid prose captures the wonder of an era in which modern astronomy was just beginning to reveal our connection to a vast universe beyond our own small world."

—John Pipkin, author of *The Blind Astronomer's Daughter*

"A suspenseful and dramatic account of the rival scientific expeditions that came to the American West to view and study this rare phenomenon. . . . Baron enables us to understand what drew them to the eclipse and what this episode tells us about the changing role of science in American culture."

—Paul Israel, author of *Edison: A Life of Invention*

"Science journalist Baron shares a timely tale of science and suspense in this story of rival Gilded Age astronomers contending with everything from cloudy skies to train robbers to observe the historic total solar eclipse of July 29, 1878. . . . Baron skillfully builds tension, giving readers a vivid sense of the excitement, hard work, and high stakes in play." —*Publishers Weekly*, starred review

"Captivating. . . . As Baron capably and enthusiastically shows, the solar eclipse of 1878 proved to be an important moment in the emergence of American science. . . . A timely, energetic combination of social and scientific history." —*Kirkus Reviews*

"Baron mingles the excitement, aspiration and drama of these events with a good dose of technical information and scientific history. Archival photos, sketches and prints are scattered throughout the pages. This is a wonderful, dramatic piece of scientific history." —Sara Catterall, *Shelf Awareness*

"In vivid detail, Baron unfolds [the scientists'] backstories and reveals what led each of them to make their way to the still unsettled Wild West to view this phenomenon. . . . *American Eclipse* will undoubtedly spur scores of readers to desire their own total solar eclipse experience." —*BookPage*

"*American Eclipse* vividly traces the journeys of three larger-than-life figures intent on making their mark during less than three minutes on that gusty July day. . . . With a wealth of choice details about their lives, Baron brilliantly presents these three pioneers, their ambitions, and their struggles." —*Booklist*, starred review

"Throughout, the book depicts the United States as a young country striving to achieve parity with Europe on the intellectual stage. . . . Baron tells a compelling tale." —*Library Journal*

"The total solar eclipse of 1878 shapes this riveting account of the rise of scientific research in the United States. . . . [Baron] perfectly captures the sense of awe one feels during a total eclipse." —Cary Seidman, *Science Teacher*

ALSO BY
DAVID BARON

———

The Beast in the Garden

HARPER'S WEEKLY.
JOURNAL OF CIVILIZATION

180.] NEW YORK, SATURDAY, AUGUST 24, 1878. [WITH A SUPPLEMENT PRICE TEN CENTS.

Entered according to Act of Congress, in the Year 1878, by Harper & Brothers, in the Office of the Librarian of Congress, at Washington.

THE GREAT SOLAR ECLIPSE.—SKETCHED AT SNAKE RIVER PASS, COLORADO, BY MR. GEORGE STANLEY.—[SEE PAGE 675.]

AMERICAN ECLIPSE

A Nation's
Epic Race to Catch
the Shadow of
the Moon and Win
the Glory of the World

David Baron

LIVERIGHT PUBLISHING
CORPORATION ·

A Division of W. W. NORTON & COMPANY

Independent Publishers Since 1923

NEW YORK LONDON

For information about permission to reproduce selections from this book,
write to Permissions, Liveright Publishing Corporation, a division of
W. W. Norton & Company, Inc., 500 Fifth Avenue, New York, NY 10110

For information about special discounts for bulk purchases, please contact
W. W. Norton Special Sales at specialsales@wwnorton.com or 800-233-4830

Manufacturing by Lakeside Book Company
Book design by Barbara M. Bachman
Production manager: Anna Oler

Library of Congress Cataloging-in-Publication Data

Names: Baron, David, 1964–
Title: American eclipse : a nation's epic race to catch the shadow of the
moon and win the glory of the world / David Baron.
Description: First edition. | New York, N.Y. : Liveright Publishing
Corporation, a division of W. W. Norton & Company, [2017] |
Includes bibliographical references and index.
Identifiers: LCCN 2017009679 | ISBN 9781631490163 (hardcover)
Subjects: LCSH: Science—United States—History—19th century. |
Science—United States—History—20th century. | United States—
Civilization—1865–1918. | Eclipses—History. | Astronomy—United States—
History—19th century. | Astronomy—United States—History—20th century.
Classification: LCC Q127.U6 B2755 2017 | DDC 523.7/80973—dc23
LC record available at https://lccn.loc.gov/2017009679

ISBN 978-1-324-09469-2 pbk.

Liveright Publishing Corporation
500 Fifth Avenue, New York, N.Y. 10110
www.wwnorton.com

W. W. Norton & Company Ltd.
15 Carlisle Street, London W1D 3BS

For my father

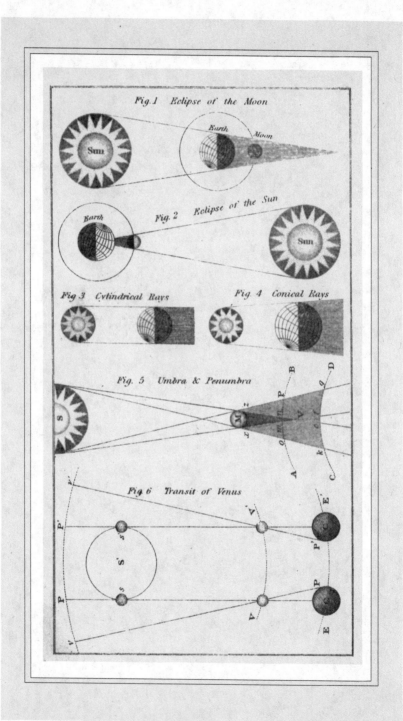

Fig. 1 Eclipse of the Moon

Fig. 2 Eclipse of the Sun

Fig 3 Cylindrical Rays

Fig. 4 Conical Rays

Fig. 5 Umbra & Penumbra

Fig 6 Transit of Venus

CONTENTS

PREFACE

To be human, it seems, is to seek purpose in our transient lives. Many people find meaning in the eyes of their children or in the words of Scripture, but I discovered it on a beach outside a Hyatt Regency in Aruba. I had journeyed south that winter of 1998 to escape the snows of Boston and, more notably, to take in nature's grandest spectacle, a total solar eclipse, which would cross the Caribbean on a Thursday afternoon in late February. As a science journalist, I thought I knew what to expect. For 174 seconds, the blue sky would blacken, stars would appear, and the sun would manifest its ethereal outer atmosphere, the solar corona. What I had not anticipated was my own intense reaction to the display.

For three glorious minutes, I felt transported to another planet, indeed to a higher plane of reality, as my consciousness departed the earth and I gaped at an alien sky. Above me, in the dim vault of the heavens, shone an incomprehensible object. It looked like an enormous wreath woven from silvery thread, and it hung suspended in the immensity of space, shimmering. As I stood transfixed by this vision, I felt something I had never experienced before—a visceral connection to the universe—and I became an *umbraphile*, an eclipse chaser, one who has since obsessively stalked the moon's shadow—across Europe, Asia, Australia—for yet a few more fleeting moments of lunar nirvana.

Over the years, this eccentric passion naturally led me to wonder how humans in the past have responded to the same imposing sight, and my curiosity eventually steered me to the Library of Congress. That institution holds not just books but also millions of artifacts culled from American history—from Lincoln's early draft of the Gettysburg Address to the correspondence of Groucho Marx—and among its vast collections are the personal papers of astronomers of the nineteenth century, the eclipse chasers of their time, who probed the hidden sun for nature's secrets. During long days at the James Madison Memorial Building, across from the U.S. Capitol, I requested box after dusty box from storage and discovered a priceless lode: faded, handwritten letters; dog-eared news clippings; telegrams and train tickets, photographs and drawings; and fragile, yellowing diaries that retained the observations, dreams, and desires of people who, like me, found magic in the shade of the moon. As I read these aging documents in the sterile glow of fluorescent lights, I grew immersed in a narrative far richer than any I had imagined. Those relics revealed a tale not just about eclipses, but about how the United States came to be the nation that it is today.

If indeed we all seek purpose in our lives, this longing applies not only to individuals but also to societies. The story I happened upon in the Library of Congress, and which I subsequently traced through archives across the continent, describes nothing less than a search for existential meaning. The tale ultimately reflects how an unfledged young nation came to embrace something much larger than itself—the enduring human quest for knowledge and truth.

Eclipses suns imply.

—EMILY DICKINSON

The Chic

THE PATH O

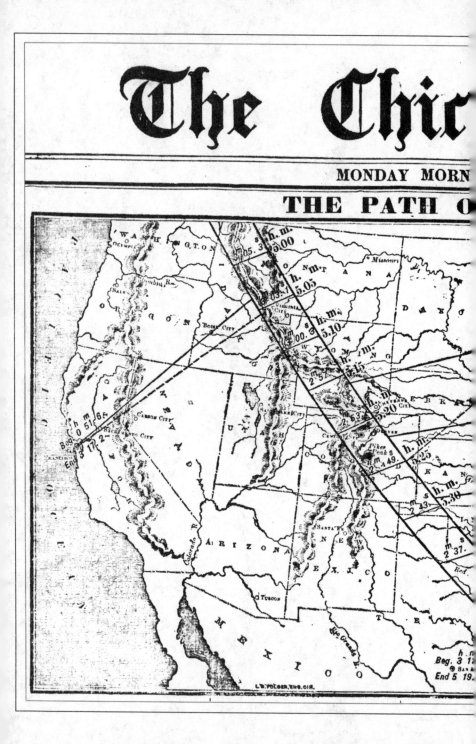

AMERICAN ECLIPSE

SHALL THE SUN BE DARKENED

JULY 29, 1878—

Johnson County, Texas

SOME WOULD CLAIM THAT THE TRAGEDY'S FATEFUL COURSE had been set several months earlier, in the winter of 1878. "That was the coldest weather I have [ever] gone against," one longtime Texan remembered, recalling that Dallas had witnessed something exotic that season: ice-skating. "They built brush and log fires all around the lake to keep from freezing while they were taking in the novel spectacle." Another local told how, after the ice had melted and the earth had thawed, the Texas summer brought another portentous phenomenon, a plague of grasshoppers. "[They] passed up nothing that was green. There were millions on millions of them." The cold snap and the locust swarms lodged in the minds of individuals who were prone to reading biblical significance into natural events. To them, these were signs that the world was soon to end.

Predictions of the world's imminent demise were not uncommon in nineteenth-century America. In the 1830s and early 1840s, followers of the millenarian preacher William Miller filled enormous tents to hear of the awesome day of Christ's return, when "the earth will be dashed to pieces" and Jesus "will destroy the bodies of the living wicked by fire." By means of an elaborate formula, Miller calculated when Judgment Day would occur, which he established as October 22, 1844—a date that became known as the Great Disappointment after the Lord did not come to retrieve his faithful, foiling the hopes of those who had climbed rooftops to prepare for their ascent into heaven. Later, the Adventist preacher Nelson H. Barbour revised Miller's calculations and offered a new forecast that he published in the book *Evidences for the Coming of the Lord in 1873*, yet another year that passed without Christ's longed-for return. By the mid-1870s, Dwight Moody, a shoe salesman turned evangelist who preached to standing-room-only crowds, steered a wiser, potentially less humiliating course. He fixed no specific date for the Rapture but implored his audiences simply to be ready at all times. "The trump of God may be sounded, for anything we know, before I finish this sermon," he intoned.

For those on the lookout for the Second Coming, celestial trumpets would not be the only harbinger of Christ's return. According to the Book of Matthew, just before Jesus appeared "in the clouds of heaven with power and great glory," another sign would manifest itself: at that moment, Christ proclaimed, "shall the sun be darkened."

ON THE SCORCHING AFTERNOON of July 29, 1878, in a landscape recently transformed from open range to farmland, the people of Johnson County were cultivating their fields. This part of North Central Texas was a mix of West and South, cowboys and cotton. The region had seen its share of slaveholding, and although the Emancipation Proclamation had ostensibly abolished the practice, for many freedmen it felt as if slavery had continued. Without property of their own, black laborers, forced to sharecrop from white landlords, still slept in rough-hewn shacks, woke before dawn, and worked interminable hours—from "can't see to can't see"—for scant reward.

What had motivated one Ephraim Miller to journey to this hardscrabble part of Texas some six months earlier is not recorded, but he had come from West Tennessee, a region that in this era saw a great exodus of "Exodusters," former slaves fleeing racial violence and economic oppression. Arriving in Texas, Miller rented a small prairie farm near the old Johnson County seat of Buchanan, with an eastward view of the Cross Timbers—dense oak woods that provided lumber for fences, plow handles, and coffins. With a wife and four children—the eldest a son, about ten—Miller was said to be prospering, at least enough to afford a recent purchase, a hatchet.

That July day had begun unremarkably, an overcast morning yielding to scattered storm clouds in the afternoon. The air was thick and hot, and lightning flashed against the summer horizon. As the sun inched westward and the hour approached four, the Texans noticed peculiarities in their surroundings. A farmer near Waco puzzled at a sight beneath his cottonwoods: the specks of light between the shadows of the leaves bizarrely turned to crescents, miniature moons dappling the ground. In Dallas, a woman on the banks of the Trinity heard the melancholy croaking of frogs. On the plains to the northwest, a nine-year-old boy caught sight of bats flying aberrantly in the afternoon. The oppressive heat began to lift as the quality of daylight shifted. The squat homes, the cornstalks, the barbed-wire fencing—everything took on an air of unreality, seemingly thrown into bold relief. The landscape dimmed—not turning gray, as if beneath cloud cover, but a faint yellow, as if lit by a fading kerosene lamp. Fireflies winked on. A star suddenly materialized, then two. The air stopped moving. The birds ceased their chatter. Then a few final ripples of light rushed over the ground—and darkness descended.

Fear swept over the fields. A man fell to his knees in supplication, between the handles of his plow. Others fled toward church. Looking up, the people of Johnson County saw an unfamiliar sky; the sun was gone, replaced by a magnificent ring of golden light—a halo. This heavenly crown was finely textured, as if made from spun silk, with hints of ruby at its base and luminous, pearly wings projecting toward the east and west.

It was then that Ephraim Miller was seen running toward home, hatchet in hand. A devout man, Miller had been heard to say that morning that he had learned the world would end that very evening, and if so, he intended to be "so sound asleep that Gabriel's trumpet wouldn't wake him." He apparently wished to avoid the apocalypse and to speed his passage to the hereafter. He did not plan to go alone. Entering the house, he encountered his son and struck hard with the axe. The boy fell, gasping for life in a pool of blood. Miller's young daughters—age two and four—wailed and hid beneath the bed, while his littlest child, an infant, crawled on the floor. Clutching a new razor with his right hand, Miller climbed a ladder to the tiny attic. There, closer to the kingdom of heaven, he cut his own throat from ear to ear. Then he fell back to earth beside his dying son.

Miller's wife, witnessing the murder-suicide, screamed and burst out the back door. "Come on, sweet chariot," she cried as she wrung her hands, crossing a cotton field in the deep twilight at the end of time.

FOR MILLENNIA, TOTAL SOLAR ECLIPSES have awed, frightened, and inspired.

In the sixth century B.C., in Asia Minor, two warring powers—the Medes and the Lydians—laid down their weapons after six years of fighting when confronted by the sudden darkness of an eclipse. (The soldiers were "zealous to make peace," Herodotus relates.) In A.D. 840, in Europe, a total eclipse so unnerved Holy Roman Emperor Louis the Pious—who had long been anxious about strange events in the heavens—that, according to an advisor, the emperor "began to waste away by refusing food" and died a month later, plunging his realm into civil war. In 1806, in North America, the appearance of a "black sun"—an omen predicted by the Shawnee prophet Tenskwatawa—emboldened a Native American uprising that his brother Tecumseh would lead against the United States in the War of 1812. The novelist James Fenimore Cooper, whose tales of wilderness adventure would captivate the nation, witnessed that same eclipse in upstate New York, and years later he recalled it vividly. "I shall only say that I have passed a varied and eventful life, that it has been my fortune to see earth, heavens, ocean, and man in most of their aspects," he wrote, "but never have I beheld any spectacle which so plainly manifested the majesty of the Creator, or so forcibly taught the lesson of humility to man as a total eclipse of the sun."

A total solar eclipse is a singular experience, not to be confused with other, more common types of eclipses. *Partial solar eclipses*, which occur at least twice a year over a large portion of the earth, offer a curious sight (through darkened glass, you can watch the moon take a bite out of the solar disk), but the effects are otherwise subtle. *Lunar eclipses*, in which the earth casts its shadow on a reddened moon, can be memorable and strangely beautiful, but they too are not especially rare. *Total solar eclipses*, on the other hand, in which the moon completely obscures the face of the sun, are exceptional— passing any given point on earth about once every four hundred years—and create an experience that is otherworldly. With a total solar eclipse, you come to appreciate that the very word—*eclipse*—is misleading, because what is notable is not what is hidden, but what is revealed. A total eclipse pulls back the curtain that is the daytime sky, exposing what is above our heads but unseen at any other time: the solar system. Suddenly, you perceive our blazing sun as never before, flanked by bright stars and planets.

Eclipses inevitably reveal much about ourselves, too. What we see in them reflects our own longings and fears, as well as our misconceptions. For Ephraim Miller, the total solar eclipse that descended over Texas on July 29, 1878, held deep religious significance, but for many others who witnessed it—especially to the north, in Wyoming and Colorado—a whole different meaning imbued the historic event. The eclipse occurred at a pivotal time in post–Civil War America. This adolescent nation, once a land of yeoman farmers, had in a mere century expanded exponentially in population, wealth, and physical extent. New technologies of the industrial age were accelerating the pace of life. Women, long confined to the home and to challenges of childbirth and child-rearing, were rebelling against cultural strictures. And now, in this consequential age of national maturation, a group of American scientists aimed to use the eclipse to show how far the country had evolved intellectually.

Although Ephraim Miller did not anticipate the eclipse, astronomers did. They computed the heavenly motions and plotted where darkness would fall, and then they endeavored to meet it, for while the event would be exceedingly brief—just three minutes in duration—it offered a chance to solve some of nature's most enduring riddles. These scientists, male and female, trekked to the western frontier in an age of train robberies and Indian wars. Bearing telescopes and wielding theories, they sought fame for themselves and glory for their country.

Among this hardy crew were a few scientists with much to prove. One astronomer was determined to find a new planet and, along with it, the acclaim he held he was due. Another meant to transform American culture by expanding the paltry opportunities for women in science. And a third, a young inventor, sought to burnish his reputation as a serious investigator, and what he learned on his journey would help him inaugurate our modern technological era. Together, these three individuals and those three minutes of midday darkness would enlighten a people and elevate a nation, spurring its rise to an honored place on the global stage.

PART ONE

1876

CHAPTER I

REIGN OF SHODDY

MONDAY, JUNE 26, 1876—

Philadelphia, Pennsylvania

EIGHT DAYS SHY OF AMERICA'S HUNDREDTH BIRTHDAY, TOURists thronged to Philadelphia for the approaching Fourth of July festivities. Elsewhere in the country, there was cause for despair. In Washington, President Ulysses S. Grant, the great Union hero of the Civil War, presided over a scandal-plagued White House. In New York, unemployment ran high among tradesmen—plumbers, shipwrights, coopers, hatters—the interminable aftereffects of the financial Panic of 1873. Far out in Montana Territory, the dashing Lieutenant Colonel George Armstrong Custer and more than two hundred of his men lay freshly slain, scalped, and dismembered above a river that the Sioux called the Greasy Grass—more familiar as the Little Bighorn—and when news of the carnage reached the East a week later, it would traumatize the nation. For now, though, in Philadelphia, a celebration was in progress. Here in the city where one hundred years earlier the country had declared its independence, Americans had come to mark the centennial at the Centennial.

Officially called the International Exhibition of Arts, Manufactures and Products of the Soil and Mine, the Centennial Exhibition was a grand world's fair that sprawled along the Schuylkill River and across the expansive Fairmount Park. More than one hundred buildings housed thousands of exhibitors from fifty nations, show-casing everything from typewriters to pianos, sewing machines to Japanese art. During the Centennial's six-month run, it would tally ten million admissions and serve as a fitting symbol for a country with world-class ambitions.

The late-June sun baked the fairground's asphalt walkways, ting-ing the air with the aroma of tar. Visitors escaped the heat by seek-ing the shade of parasols and willows, and by consuming ice cream

Machinery Hall from the Jury Pavilion

sodas and chilled drinking water, the latter available free from the Sons of Temperance fountain on Belmont Avenue, its cistern cooled by twelve hundred pounds of ice. Others sought relief in an invigorating ride on the narrow-gauge railroad that offered a five-cent, fifteen-mile-per-hour tour of the grounds—past the Women's Pavilion, Horticultural Hall, and the state buildings (New Jersey, Ohio, Indiana, Illinois), before arriving back at Machinery Hall, an imposing edifice of pine and spruce and wrought iron that spanned the equivalent of five city blocks.

Inside the vast building, fairgoers gazed at the exhibition's most impressive sight: a colossal steam engine—the world's largest—that through a massive network of shafts and gears, belts and pulleys powered eight hundred machines. With a thwack and thrum, and a whir and hiss, those machines did everything at lightning speed: spinning silk, cutting cloth, knitting socks, fashioning pins, slicing stone, planing lumber, pumping water, blending chocolate, printing and folding newspapers (at the impressive rate of 35,000 copies per hour), even weaving portraits of George Washington and Abraham Lincoln. The pulsating movements awed the crowds. The scene looked and sounded like progress, like modernity, like the nation America strove to be.

Only a decade's remove from the Civil War, the United States seemed galvanized by a surge of energy. The country was industrializing and urbanizing, laying railways and telegraph lines, settling western lands and displacing native peoples in the name of Manifest Destiny, creating a maelstrom of growth that was exhilarating for many, unsettling for everyone. It was also a time marked by the acquisition of wealth and the perceived decay of public morals, when the country's chief aim had become getting rich—"Dishonestly if we can; honestly if we must," in the words of Mark Twain. His 1873 novel *The Gilded Age*, co-written with Charles Dudley Warner, would lend its title to the era, yet that name, which has acquired a glow over time, was intended not as praise but as insult. It suggested a veneer of wealth that masked a worthless interior. Others called this period of American postwar affluence-without-culture by another pejorative: the Reign of Shoddy.

Boorish ostentation was amply on display at the Centennial, where visitors paraded in bustles and top hats while elbowing each other to view exhibits and purchase trinkets. Japan's commissioner to the exhibition described the frenzied commotion in colorful and succinct fashion. "The first day crowds come like sheep; run here, run there, run every where," he complained in *Harper's Weekly*. "All rush, push, tear, shout, make plenty noise, say damn great many times, get very tired, and go home."

ON THIS SWELTERING MONDAY at the end of June, the boisterous crowds finally did go home—chased off, as always, by a foghorn that signaled closing as the sun and temperature began to descend. In Machinery Hall, the Corliss steam engine and all that it animated now lay idle. Out the east door of the building and across a leafy quadrangle, a man as large in girth as he was in ego was about to conduct a historic test.

James Craig Watson had journeyed to the Centennial from Michigan, where he served as a professor of astronomy. A man keenly representative of his era, Watson seemed innately competitive, craving wealth and shunning scruples. He bragged of shrewd business dealings that, one fellow astronomer reflected, "seemed to me to transcend the bounds of 'sharp practice,' & run very close to fraud." He cut corners academically, failing to credit the work of colleagues—including his mentor, who complained that his former student "does not shrink from adorning himself with the merit of others." Once charged with plagiarism, Watson called the claims "trumped up" and goaded his accusers: "[A]ll I can say is they may 'fire away' for they might as well 'throw mud at a rhinoceros.'"

Watson's vanity was not without foundation, however. He had risen from an impoverished childhood of factory work and apple peddling to enter the University of Michigan as a mathematical prodigy at age fifteen. Six years later, he was on the faculty, where he proved popular with

JAMES CRAIG WATSON

students. Among the reasons, undoubtedly, was his lax grading. One year he reportedly gave passing grades on a final exam to his entire class, including a student who had died toward the beginning of the course. For his ample achievements in astronomy, Watson had earned a gold medal from the French Academy of Sciences. For his considerable rotundity, he had earned the nickname "Tubby" from his students.

In 1876, now in his late thirties, Watson received another honor, an invitation to serve as a judge at the Centennial. The great fair in Philadelphia was not just an exhibition; it was a competition in which exhibitors vied for medals bestowed by juried panels in thirty-six categories. Watson joined the jurors of Group 25—"Instruments of Precision, Research, Experiment, and Illustration, Including Telegraphy and Music"—an elite assemblage that comprised, among others, Smithsonian Secretary Joseph Henry and Britain's Sir William Thomson, a leading physicist who would later be known as Lord Kelvin.

On the previous day, June 25, Watson and his fellow judges had conducted preliminary tests of an improbable invention that purportedly could transmit the human voice by wire. Now, on this Monday evening after the fairgrounds had closed, the judges would perform a second, more thorough evaluation.

The men gathered at the Judges' Pavilion, a handsome structure capped by an arched roof and ornate towers. A wire was strung from a point outside the building to a distant room inside, the locations far enough separated that words spoken at one end of the wire could not be picked up at the other through the air. Sir William Thomson stood at the receiving end of the wire. James Craig Watson was outside, at the transmitter.

Watson clutched a copy of that day's *New-York Tribune*. He leaned in toward a vibrating membrane attached to the wire and read snippets of news items. "S. S. Cox has arrived," Watson enunciated loudly, testing sibilance. At the other end of the wire, William Thomson pressed the receiver to his ear. He discerned "has arrived" but had trouble hearing the *s*'s and *x* of "S. S. Cox." Watson tried other combinations of consonants and vowels. "City of New York." "Senator Morton." These phrases Thomson heard distinctly. Watson then tested longer strings of words. He read from an article date-lined Washington: "The Senate has passed a resolution to print one thousand extra copies. . . ." Then an item from England: "The Americans of London have made arrangements to celebrate the Fourth of July. . . ." Watson and Thomson compared the words spoken to those heard. Amazingly, the phrases matched.

These men instantly perceived that this crude device—the "speaking telephone" of one Alexander Graham Bell—would rapidly transform communications. "The results convinced both of us that Prof. Bell had made a wonderful discovery, and that its complete development would follow in the near future," Watson would later recall.

James Craig Watson marveled at other inventions at the Centennial. He and the judges of Group 25 bestowed medals on calculating machines, fire alarms, burglar alarms, gas meters, electric generators, railway signals, and a clever device being demonstrated in the exhibition's Main Building, near the Gatling gun display. This was the electric pen, an aid to the businessman that enabled quick production and reproduction of written documents. The tip of the battery-powered "pen" consisted of a fine needle that darted up and down, fifty times per second, leaving a trail of perforations when traced across paper. "It is very much like holding," one journalist wrote, "the 'business end' of a wasp on a sheet of paper and letting the insect sting small holes into the sheet while you

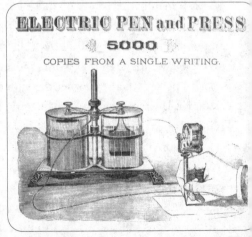

ELECTRIC PEN and PRESS
5000
COPIES FROM A SINGLE WRITING.

move him back and forward." In this way, the electric pen created a stencil, which—when placed in a duplicating press and inked by a roller—could generate hundreds of copies. "The simplicity of the whole apparatus, and the results obtained by it, entitle it to a place among the really useful inventions of the age," the judges wrote in their report—lifting the language verbatim from the company's marketing materials. "The apparatus is the invention of Thomas A. Edison."

THOMAS ALVA EDISON, AGE twenty-nine in June of 1876, was not the celebrity he would soon become, but he had already achieved considerable success as an inventor, holding more than one hundred patents. Though born in Ohio, Edison, like James Craig Watson, grew up humbly in Michigan. Unlike the astronomer, however, Edison did not seek a better life through academia. He was home-schooled and self-taught. As a boy, he sold newspapers on the Grand Trunk Railroad between his home, Port Huron, and Detroit, then spent his early adulthood not in college but as an itinerant "lightning jerker"—a telegraph operator—an occupation that took him to six states in as many years. When not sending or receiving messages down the wire, he dreamed up ways to improve the telegraph equipment and honed his inventive skills by rigging devices to deliver mild shocks (as a joke) to his fellow telegraphers, and lethal ones to the rats and insects that infested the squalid offices. "[A] flash of light and the cockroach went into gas," he recalled proudly of an early invention.

As he grew older, Edison retained his boyishness—among his most conspicuous qualities were a folksy demeanor and disheveled hair—but he matured professionally. From his base of operations in Newark, New Jersey, he invented not only the newfangled electric pen but also an improved stock ticker, which would be widely used by Wall Street, and devices that boosted the efficiency and profits of the telegraph industry. One such invention, the quadruplex, enabled

THOMAS A. EDISON

four telegrams to be sent simultaneously over a single wire. Another, the automatic telegraph, increased the speed at which messages could be sent. Edison received handsome royalties for these devices, and he earned praise from the judges at the Centennial, who called his automatic telegraph "a very important step in land-telegraphy."

Edison, Bell, and other inventors impressed the world with American know-how at the exhibition in Philadelphia. A Swiss watchmaker who served as one of the judges returned home to report on the young nation's technological prowess. "Up to this very day we have believed America to be dependent upon Europe. We have been mistaken," he warned his countrymen. "We have believed ourselves masters of the situation, when we really have been on a volcano." But for all of America's inventiveness—its "knackiness," in the words of a British visitor—the Centennial also showed where the nation lagged. The United States was good at making gadgets and money, but what else?

Foreigners scoffed at displays of American art. American music was held in such low regard that the Women's Centennial Committee, seeking a composer to write the exhibition's inaugural march, hired a German—the commanding and megalomaniacal Richard Wagner. (The march proved a disappointment nonetheless.) America's reputation in the sciences, such as physics and chemistry, also fared poorly (although it fared somewhat better in geology and paleontology, thanks to a surge of discoveries in the newly explored reaches of the trans-Mississippi West). Philadelphia industrialist Joseph Wharton lamented that the Centennial revealed a "rather meagre display of American scientific achievement."

This lack of achievement did not, however, reflect a lack of will. By 1876, a generation of American intellectuals, acutely sensing their own inferiority as they gazed across the Atlantic, had striven to elevate their country to scientific greatness. They had founded scientific societies and associations, institutes and academies, schools, lyceums, and libraries. Still, where Europe could boast many scientific luminaries such as Joule and Ampère and Gauss—names so esteemed they would soon be immortalized as units of measurement—the United States could claim a pittance. (One exception was the Smithsonian's Joseph Henry, for whom the henry—a unit of electrical inductance—would later be named.) In the 1870s, American scientists often went to Europe for graduate education, and, once there, they found their homeland disparaged. The French "think we are a mere nation of moneybags and insignificant students," complained a leading American astronomer, Simon Newcomb. And it was not just Europeans who criticized the precarious condition of science in the United States. Newcomb was among several prominent Americans who engaged in public self-flagellation. At the beginning of 1876, asked to review the progress of American science over the previous century, Newcomb penned a harsh critique, pointing to a "period of apparent intellectual darkness" that had only begun to lift.

No one, however, articulated the intellectual challenge to America as forcefully as Thomas Henry Huxley. The British naturalist, who was famous—and infamous—for his fierce defense of Darwinism, visited the United States on a lecture tour in 1876. Before stopping at the Centennial, in Philadelphia, he gave the inaugural address at the new Johns Hopkins University, in Baltimore. Huxley stressed the importance of scientific research to the progress of humankind, and he spoke bluntly to this striving, expanding, adolescent nation. "I cannot say that I am in the slightest degree impressed by your bigness, or your material resources, as such," he proclaimed in a passage that would be widely quoted. "Size is not grandeur, and territory does not make a nation. The great issue, about which hangs a true sublimity, and the terror of overhanging fate, is what are you going to do with all these things? What is to be the end to which these are to be the means?"

IN EARLY JULY 1876, just after the United States celebrated its hundredth birthday and as the Independence Day festivities died down, a London scientific journal published a small item in its astronomical column. The article in *Nature* observed that in two years' time, a total solar eclipse would pass over America's western states and territories. The article provided details of where the moon's shadow would fall, and it included a statement that might be interpreted as encouragement or challenge. Of the eclipse, the British journal wrote simply, "Our American *confrères* will no doubt give a good account of it."

CHAPTER 2

PROFESSOR OF QUADRUPLICITY

JULY 1876—

Menlo Park, New Jersey

After six hectic years in Newark—opening factories, fighting lawsuits, fending off creditors—Thomas Edison sought an escape in the rolling green countryside south of the city. For $5,200 he bought two land parcels, one with a house, the other on which he built a workshop. In a letter to a friend, Edison waxed poetic about the new laboratory at "Menlo Park, Middlesex Co., New Jersey, U. S. A., Western Div. Globe, Planet Earth, 4 miles south of Rahway on Penn R. R. on a High Hill." He called it "the prettiest spot in N. J." and added, "will show you round—go strawberrying."

Edison's new laboratory stood conveniently two blocks west of

GENERAL VIEW OF MENLO PARK AND EDISON'S LABORATORY.

INTERIOR OF EDISON'S MACHINE SHOP WHERE HIS EXPERIMENTS ARE CONDUCTED.

the railroad tracks. Twenty-five feet wide by one hundred long, the building "in size and external appearance, resembles a country church," one visitor commented. "The interior, however, is not so church-like." The downstairs machine shop recalled the Centennial's Machinery Hall: aproned men at steam-powered lathes and drills sculpted raw metal into whatever Edison envisioned. The large upstairs room looked more like an apothecary, its floor-to-ceiling shelves presenting more than a thousand glass jars labeled with their contents: acids, salts, oils, and plant extracts, including a red sap called "dragon's blood." From the laboratory's second-floor balcony, on a clear day, one could make out the towers of the new bridge rising across the East River, twenty miles away, that would eventually link New York to Brooklyn.

Just down the hill, in a gabled dwelling trimmed an earthy red, Edison's young wife presided. Mary Edison decorated the family home with bronzes and busts and a pianoforte. She threw parties and tended to the children: Thomas Alva Jr., born in January 1876, and three-year-old Marion Estelle, the pair nicknamed Dot and

Dash by their telegrapher father. The Edisons' marriage was typical of a certain class of the era. She took care of domestic affairs, albeit with the help of servants. He looked after business. "My Wife Popsy Wopsy Can't Invent," he scribbled in a technical notebook one Valentine's Day.

Edison spent all hours with his fraternity of workmen up at the laboratory, covered in grease, and often did not return home until dawn. "I like it first-rate out here in the green country and can study, work and think," he said. Mary felt neglected. She feared burglars. She slept with a revolver under her pillow.

EDISON'S MENLO PARK "INVENTION FACTORY" would become legendary. It was here that, during a few years of frantic activity, he would conceive his most famous inventions, in addition to a host of others long since forgotten. For each of Edison's celebrated triumphs, he experienced many defeats, and this has led some to conclude that

EDISON'S HOME, MENLO PARK, NEW JERSEY.

Edison's accomplishments stemmed not from intelligence but from sheer persistence, perhaps mania—that he was a lucky tinkerer who succeeded simply by trial and error.

Indeed, Edison promoted this myth. He publicly disdained intellectualism. "I wouldn't give a penny for the ordinary college graduate, except those from the institutes of technology," he harrumphed. "They aren't filled up with Latin, philosophy, and the rest of that ninny stuff." Once, when a news article referred to Edison as a "scientist," he protested. "That's wrong! I'm not a scientist," he insisted. "I'm an inventor." But Edison made these comments later in life, after a divide, indeed a chasm, had opened between him and the scientific establishment. The Thomas Edison who moved to Menlo Park was not dismissive of scientists. In fact, he aspired to be one.

Edison may have lacked formal schooling, but he did not lack science education. He had read Isaac Newton, albeit with difficulty, by age twelve; devoured physics books throughout his teens; and, in his twenties, employed a Brooklyn high school teacher to tutor him in chemistry and acoustics. At Menlo Park, Edison's library included the latest scientific journals (as well as the poetry of Poe and the ever-popular Longfellow). Edison also befriended some of the era's top academic scientists. Sir William Thomson, after serving as a judge at the Centennial, stopped at Menlo Park before returning to England. Professors from Princeton were frequent visitors.

Edison's closest associate from academia was George F. Barker. "I am the Professor of Physics in the University of Pennsylvania," Barker wrote with formality in 1874, inviting Edison to Philadelphia. "[I] shall be glad to see you [at

GEORGE F. BARKER

the university] when you come; or at my house." Edison obliged, and the men struck up a lasting relationship. Barker, a bespectacled man with the whiskers of a schnauzer, gave popular lectures that wowed audiences with demonstrations of electrical science, often presented from the same Philadelphia stage that featured Italian opera and hosted lectures by Ralph Waldo Emerson. During the Centennial, Barker had helped arrange the first demonstration of Alexander Graham Bell's telephone, and he was now keenly interested in Edison's attempts to build a rival telephone. The distinguished physics professor would become one of Edison's greatest champions.

If Edison sought to emulate any professor, however, it was a man no longer living. Michael Faraday, chemist and physicist at Britain's Royal Institution, had transformed scientific understanding of electricity and magnetism by showing that those two natural forces, seemingly distinct, are bound up as one. During years of clever, meticulous study, largely in the 1830s, Faraday demonstrated that magnets could induce electric currents and electric currents could produce magnetism, and he thereby invented the basic elements of the electric motor and the electric generator. A humble, hands-on investigator who was not above placing electrodes on his tongue to test for electricity, Faraday believed a true scientist should question everything, including oneself. "The man who is certain he is right is almost sure to be wrong, and he has the additional misfortune of inevitably remaining so," he once said. "All our theories are fixed upon uncertain data, and all of them want alteration and support."

Faraday published detailed accounts of his investigations in a massive, three-volume work called *Experimental Researches in Electricity*. Thomas Edison bought and studied it. "I think I must have tried about everything in those books," Edison commented. "His explanations were simple. He used no mathematics. He was the Master Experimenter." By autumn 1875, Edison was modeling his scientific work after Faraday—in fact, Edison labeled his laboratory notebook "Experimental Researches"—and he immediately believed he had made a discovery worthy of his scientific idol.

In late November of that year, while experimenting with new modes of telegraphy, Edison constructed a device that vibrated rapidly by means of an electromagnet. The contraption clicked like a hyperactive telegraph key, and as it did so, it threw off a burst of bright sparks, including from metal objects near the device yet disconnected from it. Intrigued, Edison performed tests on those sparks. He and his assistants touched their tongues to metal but could feel no electric shock. The men tried to measure electric current by means of a *galvanoscopic frog*—an amputated frog's leg with its sciatic nerve exposed, a sensitive detector also used by Faraday—but again the sparks triggered no response. Edison concluded that these sparks did not display the properties of electricity and must therefore represent something entirely new. Far less circumspect than his British hero, Edison quickly announced his findings to the newspapers.

"Mr. Edison . . . promises to become famous as the discoverer of a new natural force," blared *The New York Herald*. The *New-York Tribune* expounded, "Mr. Edison has named the new principle 'etheric force,'" which *The Daily Graphic* described as "a sort of first cousin to electricity" and which *The Boston Globe* opined "may prove of great value to the science of telegraphy." Indeed, Edison believed he had made a fundamental discovery in physics, one that might also revolutionize communications, for it seemed that his etheric force was transmitted more easily than electric currents.

Edison had, in fact, stumbled on something revolutionary—without realizing it, he had discovered radio waves more than a decade before German physicist Heinrich Hertz would reveal their true nature—but Edison was wrong to assume that what he had observed was a new natural force, and his hasty conclusion soon drew the scorn of scientists and engineers. William Sawyer, an electrical inventor and one of Edison's chief rivals, wrote harshly,

"I do not hesitate to pronounce the whole thing, both as concerns the public and in a scientific point of view, as one of the flimsiest of illusions." *Scientific American* concluded that Edison's declaration of a new force had been "simply gratuitous."

The humiliation dogged Edison even after he moved to Menlo Park. James Ashley, editor of a professional journal called *The Telegrapher*, had at one time been a business partner of Edison's, but the partnership did not end well, and Ashley often retaliated in his journal; he printed snide articles that disparaged Edison's ethics and belittled his inventions, including the quadruplex telegraph. In July of 1876, now that Edison's foray into scientific research had become an embarrassment, Ashley wrote another article drenched in sarcasm:

Edison About to Astonish the World Again.—Stand from Under!

The professor of duplicity and quadruplicity has been suspiciously quiet for some time. Since his great discovery of the new moonshine, which he christened "etheric force," he has apparently subsided. . . . Satisfied that some great purpose was concealed under this reticence, and determined that the world, and especially the telegraphic world, should not remain in ignorance of the doings of the most remarkable genius of this or any other age and country, THE TELEGRAPHER has taken the trouble to penetrate the mystery which enshrouds his purpose. It has been discovered that the professor is about to astonish the world, and confound the ignoramuses who are engaged in the improvement of telegraphic apparatus, by the production of an invention which has taxed his massive intellect and unparalleled inventive genius to the utmost. . . . The establishment at Menlo Park has not been created for nothing.

That secluded precinct is yet to become famous throughout the earth as the spot where this invention was conceived and brought to light!

Edison had attempted to prove himself a respected scientific investigator, in the mold of Michael Faraday. He had, instead, been taken for a fraud and a fool.

Yet the "professor of duplicity and quadruplicity" was not easily dissuaded. Tapping a deep reservoir of persistence and self-confidence, he would indeed try again to astonish the world, to demonstrate that he was a scientist and no mere tinkerer.

CHAPTER 3

NEMESIS

WEDNESDAY, SEPT. 27, 1876—

Ann Arbor, Michigan

JAMES CRAIG WATSON, THE ASTRONOMER WHO HAD JUDGED Alexander Graham Bell's telephone at the Centennial Exhibition, lived and worked in his own pastoral setting, at the University of Michigan's Detroit Observatory. Set on a grassy ridge a half mile northeast of campus, the building was named not for its

OBSERVATORY OF MICHIGAN UNIVERSITY.

location—Ann Arbor—but to honor the city in which its benefactors lived. The edifice stood as a monument to philanthropy and science. Its bracketed cornice and Doric portico lent a Greek Revival elegance to the exterior, while its all-seeing eye lay beneath the domed roof. From the observatory atop the hill, one could look down to the Huron River and up toward eternity.

It was a Wednesday night at the start of the school year, and Watson was at the great refracting telescope, an instrument so large it dwarfed even a man of such generous proportions. The wooden tube, more than seventeen feet in length, was connected to a clockwork that kept it aligned with the heavens. The telescope was mounted equatorially—along the plane of the equator—and as the earth slowly turned, it remained fixed against the stars. The whole contraption sat on its own pier, a masonry foundation that descended forty feet through the center of the building and another fifteen feet underground, isolating the telescope from footsteps and other sources of vibration. Watson pulled a thick rope to rotate the dome, which rolled on cannonballs, and he aimed the large telescope through the narrow, shuttered opening. Cool autumn air rushed in along with the sounds of the night.

These were Watson's prime working hours. While his wife slept in the expansive residence next door, he scoured the Michigan sky. "He knew the stars as one knows the faces of his friends," a colleague recalled, but what Watson most sought were strangers, and he displayed a great talent for finding them. As a newspaper described it: "Two or three or four times a year he turns up with a pocket full of new planets that he has picked up as a boy does marbles."

Indeed, Watson was, in the parlance of the day, a planet hunter, and on this night he found his quarry—a dim, shining object in Pisces. Almost immediately, he telegraphed his discovery to Joseph Henry at the Smithsonian, who then wired the news to Europe. It was, remarkably, Watson's nineteenth planet, and he would name it Sibylla.

———

IN WATSON'S ERA, the term *planet* commonly denoted two very different classes of objects. Mercury, Venus, Earth, and the other large bodies that orbit the sun at widely spaced intervals bore the label *major* planets, while vastly smaller objects—"so small that a good walker could easily make the circuit of one of these microscopic globes in a single day," as one scientist put it—were *minor* planets, a term still used, in fact, among astronomers. Rarely visible to the naked eye and mostly clustered in a belt between Mars and Jupiter, these tiny worlds are better known as asteroids.

So many asteroids have now been identified that they have been famously disparaged as "vermin of the skies," but in the 1870s they were still relatively novel, and a heated international race was underway to see which country could claim the largest share. Astronomers found the first minor planet at the very inception of the nineteenth century, and fifty years later—when the rate of discovery accelerated—just a dozen had been added to the list. Finding these diminutive worlds required mind-numbing labor: concentrating on a small patch of sky, mapping all stars visible through one's telescope, and reexamining that same region night after night until a new starlike object appeared that moved almost imperceptibly across the field. The discoverer then had to record the asteroid's exact coordinates and calculate its orbit so it could be tracked and verified as a new object, not simply a rediscovered old one. Hunting planets mandated machinelike precision and an obsessive drive—traits exhibited by James Craig Watson even at the beginning of his career.

At age nineteen, while serving as an assistant at the Detroit Observatory, Watson had espied a tiny dot moving through the heavens. By that time—the fall of 1857—fifty minor planets had been found, yet only two by Americans. News of Watson's asteroid reached Christian Heinrich Friedrich Peters, an older astronomer of Danish birth and German heritage who had recently immigrated to the United

C. H. F. Peters

States and would soon head the observatory at Hamilton College in the upstate town of Clinton, New York. "[W]e have great hope that yours is a new one," Peters wrote to Watson about his planet. "[E]very discovery in this country must be of double interest to us, in order to teach old-foggy [*sic*] Europe to look a little more around towards this side of the Atlantic." Alas, it turned out that Watson's planet had been seen a few weeks earlier in Düsseldorf, so credit for the discovery was assigned, disappointingly, to the Old World. Watson wrote back to Peters, his new compatriot. "I wish they would stop awhile in Europe, and we would have a better chance of bringing up the rear, and of commanding the advance."

Before long, C. H. F. Peters and James Craig Watson did command the advance. The American public at the time was absorbed in more pressing matters—armed conflict had severed the nation—but in 1861, just a month after the Civil War began, Peters found his first asteroid. He called it Feronia, for the Roman goddess of groves and freedmen. (Like the major planets, minor ones received names, customarily those of mythological figures.) The following year, in the autumn, shortly after the horror that was the Battle of Antietam, Peters found two more minor planets, Eurydice and Frigga. Watson entered the race in 1863 with Eurynome.

Come 1865, with the war finally over but the nation now grieving its assassinated president and the loss of a generation of young men, the two scientists continued to collect planets, and they occasionally stepped on each other's toes. In October, Watson announced that he had discovered an asteroid only to learn that Peters had noticed the same object a few weeks earlier. "I can surrender my claims to this one with becoming grace," Watson wrote with uncharacteristic humility. Later, when Peters inadvertently claimed one of Watson's recent discoveries, the Hamilton College student newspaper reported, "Dr. PETERS gracefully yields the palm." It was all very cordial. Together, these men reflected glory on their now reunified country, and Peters, the more experienced of the pair, maintained a respectable lead over Watson, his younger colleague.

Then came Watson's *annus mirabilis*.

Up through 1867, no astronomer—let alone an American—had discovered more than four asteroids in a single year, yet in 1868, Watson found Hecate, Helena, Hera, Clymene, Artemis, and Dione, a yield of half a dozen. "I congratulate you upon the discovery of your new asteroids," Peters wrote to Watson, "it seems you have fallen upon a nest of them." Watson's home institution, the University of Michigan, was less magnanimous in its response. "The score stands . . . Watson 9, and Peters 8," a student magazine gloated. "The fact is, that Prof. Watson has discovered more asteroids than any other man in America." What began as a friendly team sport—the United States against Europe—became a personal competition between the two men, a battle that grew increasingly bitter and divisive.

That these two astronomers would clash was probably inevitable. If Watson was headstrong, Peters was irascible. "Of his personality it may be said that it was extremely agreeable so long as no important differences arose," Simon Newcomb recalled. And when differences did arise, Peters relished tweaking his adversaries. Such was the story behind one of Peters's asteroids, Miriam. He named the planet not after a Roman or Greek goddess, but after the sister of Moses, and he did it, a fellow astronomer complained, "in defiance of rule, and of malice aforethought; so that he could tell a theological professor, whom he thought to be too pious, that *Miriam*, also, was 'a mythological personage.'" Peters vehemently disliked pomposity. It is no surprise then that he came to dislike Watson.

Through the early 1870s, as the nation's Reconstruction sputtered and its industry began to soar, the rivalry between these men only intensified. At his observatory in upstate New York, Peters claimed Felicitas, Iphigenia, Cassandra, Sirona, Gerda, Brunhilda, Antigone, Electra. From Ann Arbor, Watson snagged Althaea, Hermione, Aethra, Cyrene, Nemesis.

A Detroit newspaper, siding predictably with Watson, the local competitor, called the contest a "great planet shooting match, in which [the astronomers] are engaged for the belt (Orion's) and the championship of the globe." A daily in Atlanta marveled, "If these fellows go on picking up loose stars in this way they will soon have the planetary system so crowded that a new one will have to be invented." A New York tabloid was more jaded: "The picking up of asteroids by Watson and Peters is . . . becoming monotonous."

By the time Watson found Sibylla in September 1876, only two months after the nation's hundredth birthday, he and Peters had propelled their upstart country to the front of the international planet-hunting race, the United States having pulled ahead of France by a score of fifty-one to forty-nine. But Watson had conspicuously lost ground to his rival. In the individual rankings, Peters had taken a commanding lead, twenty-six to nineteen. An Ann Arbor newspaper was left to concede that its hometown contestant was losing the race. "Prof. P. is beating Prof. Watson 'clean out o' sight.'"

Constitutionally obsessive, Watson would not give up so easily. He was, as a colleague remarked, "extremely self confident" and "selfish and unscrupulous in advancing his own interests." A clue to his personality can be found in a notebook he kept as a college junior. Within the more than two hundred pages of lecture notes, mathematical formulae, and designs for telescopes, he continually practiced his signature—as if perfecting his autograph—and at one point he seemed to imagine his epitaph. "The Hon. James C. Watson one of the greatest astronomers that this Country has produced to whom unmeasured devotion to science owes some of its greatest blessings," he scribbled self-referentially. "Astronomy under his patronage has reached a summit rarely attained."

Watson pictured a glorious future for himself and, through his achievements, for a grateful nation. Although by 1876 his competitor had bested him, Watson saw a way to regain the edge. He would hunt new and larger game, not only seeking minor planets, but doggedly pursuing a major one.

CHAPTER 4

"PETTICOAT PARLIAMENT"

WEDNESDAY, OCTOBER 4, 1876—

Philadelphia, Pennsylvania

THE AFFAIRS OF MONEY AND MEN DOMINATED THE HEADLINES of America's Gilded Age, as was evident in the pages of the Philadelphia dailies on Wednesday, October 4, 1876. Along with the latest stock quotes and baseball scores and the price of whale oil and molasses, there was ample news of politics, an almost exclusively male domain in the era before women's suffrage. Colorado had just achieved statehood and held its first elections. "Large Republican Gains in the Centennial State," crowed the front of *The North American*. *The Press* considered another topic governed by men: the ongoing conflict with America's beleaguered Indians. "I think that the war is practically over now, and that Sitting Bull will never be able to gather such a large force again as that with which he crushed Custer," it quoted John B. Omohundro, a famous frontiersman better known as "Texas Jack." Closer to home and considerably less harrowing was news from the Centennial grounds in *The Philadelphia Inquirer*. "The principal event of yesterday was the Bankers' Convention, the proceedings of which are herewith appended," it wrote.

Elsewhere in the city, another convention aimed to make news of a different sort. As a light rain fell, a crowd entered a white marble building at Thirteenth and Arch, then ascended a broad stairway to an elegant hall of pillars and frescoes. The space was owned and rented out by an English aid society that served as a cultural bridge between Britain and America. Painted medallions of George Washington and Benjamin Franklin gazed down from the ceiling, while a full-length portrait of Queen Victoria, soon to begin the fifth decade of her long reign, graced the far end of the hall. The painting depicted the monarch in her younger, trimmer days, the vermilion background catching the rouge in her cheeks. The image evoked femininity and fortitude—an appropriate symbol for this gathering.

The occasion was the Woman's Congress, an annual symposium that explored social issues from a female perspective. It attracted a large crowd, almost entirely women, consisting of artists and doctors, suffragists and teachers, journalists and poets—including Julia Ward Howe, famous for her "Battle Hymn of the Republic." And presiding over it all, up on the platform adorned with flags and flowers, was an astronomer in a dress of black silk. She stepped to a desk, called for a moment of silent prayer, and opened the proceedings with a speech.

"When we inquire in regard to the opportunities afforded to women for the study of science, we are not surprised to find them meager and unsatisfactory," Maria Mitchell began. "Taking our whole country into consideration, there is very little attention paid to science. The same influences which deter men in scientific research operate only more forcibly upon women," she said. And then she made what many in American society would consider a bold, even distasteful, proposal. "I should like to urge upon young women a course of solid scientific study in some one direction for two reasons. First: the needs of science. Second: their own needs."

AT A TIME WHEN almost all professional scientists were male, and defiantly so, Mitchell (whose first name was pronounced muh-*rye*-uh) served as America's most notable exception. She had gained renown in 1847 when, while working as a librarian by day and studying the

heavens by night, she discovered a comet—a feat that
earned her a gold medal
from the king of Denmark
and, perhaps more remark-
ably, a paying job with a
branch of the U.S. Navy.

In that era, safe navigation
of the world's oceans—and
therefore the very efficiency
of global trade—depended
on a keen understanding of
the celestial spheres. "Astronomy enters into the price of
every pound of sugar, every
cup of coffee, every spoon-
ful of tea," the U.S. Naval
Observatory asserted. Mariners tracked latitude and longitude by
carefully measuring the positions of the sun, moon, stars, and planets and comparing those readings to an astronomical table, called an
ephemeris, which served like a railroad schedule of the skies, specifying the daily movements of the heavenly bodies months in advance.
Even small inaccuracies in these tables could send ships dangerously
astray, so the Navy employed expert astronomers and mathematicians to plot precise orbits well into the future. Such calculations
had to be done by hand, by workmen called *computers*, who penned
rows and columns of tiny numerals in oversized ledgers, a job that
demanded rare skill, concentration, and stamina.

Lieutenant Charles Henry Davis, superintendent of the Navy's
Nautical Almanac Office, which published the American ephemeris, had hired Maria Mitchell to do the computations for Venus—
an appropriate assignment, he gallantly argued, since the planet
was named for the Roman goddess of beauty. ("As it is 'Venus who
brings everything that's *fair*,'" he wrote to Mitchell in language that

typified the time, "I therefore assign you the ephemeris of Venus—you being my only *fair* assistant.") Mitchell did this work for nineteen years, even after she took a second job, as the first professor of astronomy at a new, all-women's college founded by Matthew Vassar in Poughkeepsie, New York.

By the time of the Woman's Congress in Philadelphia, Mitchell had been teaching at Vassar for a decade and was approaching sixty, but while her curls had grayed, the lessons of childhood still ran fresh in her veins. A daughter of Quakers, she had grown up on Nantucket in a community that embraced women's education. She also claimed a hereditary love of science, as she was a distant cousin of Benjamin Franklin. "The lightning that he caught on the point of his kite seems to have affected the whole race," she wrote. Her father, an amateur astronomer, encouraged Maria's independent study from the family's rooftop observatory, and he enlisted her at age twelve to help him observe a solar eclipse that passed over their island home. (That event, which occurred on February 12, 1831, was an *annular* eclipse, in which the moon passes directly in front of the sun but is too far from the earth in its orbit—and therefore appears too small in the sky—to cover the solar surface entirely. At the peak of such an eclipse, the land does not go dark but the sun is left a luminous ring, or *annulus*, in the sky.)

As Mitchell matured and ventured beyond Nantucket, she came

VASSAR COLLEGE.

to understand—and increasingly resent—the second-class status to which women were relegated in the sciences. In her late thirties, Mitchell toured Europe for almost a year, as was a custom of the educated and affluent. When not sightseeing with Nathaniel Hawthorne's family and teaching his children the constellations (Hawthorne's young son, Julian, learned to identify the bright stars in what he called "O'Brian's Belt"), she visited the great Old World observatories, where she quickly found that the men in charge did not always treat a "lady" astronomer as a peer. In France, the head of the Paris Observatory invited her to tea and gave her a cursory tour, yet he declined to show her the domes. "[I]t was evident he did not expect me to understand an observatory," Mitchell vented. In Rome, although she received admittance, after special pleading, to the Jesuit observatory atop the Church of St. Ignatius—which was generally off limits to women—the astronomer-priest in charge denied her request to remain for nighttime viewing. "[T]he Father kindly informed me that my permission did not extend beyond the daylight."

Later, back in America, Mitchell faced unequal treatment at Vassar. Despite teaching at a women's college, she received less than half the salary paid to the school's male professors—an injustice she fought, with some success—and while astronomers at other universities (for instance, James Craig Watson at Michigan) were offered generous faculty housing, Mitchell occupied the sparest of accommodations. For her first decade at Vassar, she slept on a sofa in the corner of a box-like space that alternately served as parlor and lecture room. When the college finally provided her with a separate apartment for sleeping—it had previously been the observatory's coal storeroom—one of her students marked the occasion in tongue-in-cheek verse:

Beautiful Venus, pride of the morning,
Tell it to all little stars who have fled
That, in a sweet chamber that needs no adorning,
Miss Mitchell sleeps in a bed.

Despite her clashes with the college administration, Mitchell loved her "girls," as she called her students, and in turn they adored her. It was for them that she took up arms in the fight for women's higher education, a campaign she waged even as a storm of opposition billowed.

IN 1873, A YEAR WHEN the nation was grappling with financial panic, a prominent Boston physician introduced a new cause for public alarm. In an incendiary book called *Sex in Education; or, A Fair Chance for the Girls*, Dr. Edward H. Clarke warned that the push for female colleges and coeducation could seriously undermine the health of American women. He contended that by taxing the brain, higher education caused a girl's body—especially her reproductive organs—to atrophy. "When arrested development of the reproductive system is nearly or quite complete, it produces a change in [a woman's] character," he wrote, and this included "a dropping out of maternal instincts, and an appearance of Amazonian coarseness and force. Such persons are analogous to the sexless class of termites." Clarke recounted case studies of previously healthy girls who, after studying at Vassar and other schools, became pale, sterile invalids. In one instance he diagnosed "death from over-work," and a new term, "Vassar victims," entered the late nineteenth-century lexicon.

Thrown on the defensive, proponents of women's education published a barrage of rebuttals, offering evidence of the health of college girls and stressing the benefits of higher learning. "If we know the number of young girls who have died from over study, let us find the number who have died from aimless lives, and the number who have lived on and ceased to be young," said the redoubtable Maria Mitchell. Clarke's critics called out his book for what it was: a thin, hysterical polemic, based on conjecture and scanty evidence.

No matter. *Sex in Education* found a receptive audience in an America unsettled by shifting gender roles. The Civil War, having tragically killed well over half a million men, left many women unmarried, forcing them to enter the workforce. Meanwhile, the successful effort to abolish slavery had inspired women to seek their own full citizenship, including the right to vote, which was no longer denied—at least in theory—to freed black men. American society was changing so irrevocably that it seemed women were in danger of no longer being women, and men would soon be emasculated and cease to be men. Clarke's book foretold this horrid future.

Gatherings such as the Woman's Congress stoked these public fears. During the three-day proceedings in Philadelphia, one ardent speaker would boldly assert the "full and positive equality, physical and mental," between men and women. Another would rebuke the mores of extramarital sex that "crush and dishonor the offending woman" yet "mildly admonish the offending man." A third would call for the reform of women's dress and its "tatterdom of flimsy, frayable, soilable outrigging that makes us unfit for efficient work or comfortable play." The general public's response to such remarks was often hostile. Newspapers lampooned feminists as jilted man-haters, a caricature that *The Chicago Times* had applied to the Woman's Congress in 1875. "Here were assembled all those blighted hearts whom vile men refuse to take as partners of their joy and sorrow. All who have, for cause or without, quit the bed and board of abandoned tyrants unfit for association with the gentler sex." Another newspaper ridiculed the gathering in Victorian idiom as a "petticoat parliament."

It was against this backdrop that Maria Mitchell—herself an unmarried professional—presided at the Woman's Congress in 1876. But Mitchell was no strident agitator. Shy by nature, she took to public speaking reluctantly, and as she declaimed from the podium and encouraged young women to follow her into a life of science, she carefully staked out a middle ground, assuaging the public fear and prejudice that served as obstacles. She chose not to argue that men and women were the same. In fact, she took the contrary position.

"Women are needed in scientific work for the very reason that a woman's method is different from that of a man," Mitchell said. "All her nice perceptions of minute details, all her delicate observation of color, of form, of shape, of change, and her capability of patient routine, would be of immense value in the collection of scientific facts. When I see a woman put an exquisitely fine needle at exactly the same distance from the last stitch at which that last stitch was from its predecessor, I think what a capacity she has for astronomical observations. Unknowingly, she is using a micrometer [to measure the angles between stars]; unconsciously, she is graduating circles. And the eye which has been trained in the matching of worsteds is specially fitted for the use of prism and spectroscope."

Having come of age in a brash, misogynistic world, Mitchell must have understood the futility of challenging men too directly. She was arguing merely for equality of opportunity. "Does anyone suppose that any woman in all the ages has had a fair chance to show what she could do in science?" she asked with a ring of protest in her voice.

Within the confines of St. George's Hall, Mitchell was speaking to an audience of believers. It was America at large that needed to hear her message, to be convinced that science and higher learning were not anathema to femininity, and that women deserved a fair chance.

Although the hall suffered from poor acoustics and the audience had difficulty hearing the speech, Mitchell's words resonated well beyond Philadelphia. A few weeks later, a prominent newspaper far to the west printed the lecture under the title: "The Need of Women in Science." The publication was the *Rocky Mountain News*, in the Centennial State of Colorado, out on America's frontier, where the nation was in so many ways reinventing itself.

PART TWO

1878

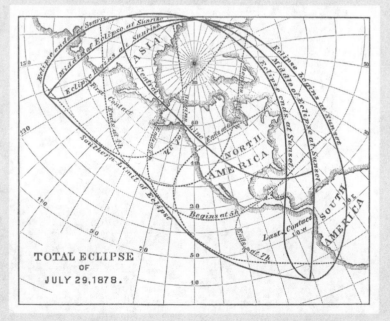

TOTAL ECLIPSE
OF
JULY 29, 1878.

CHAPTER 5

POLITICS AND MOONSHINE

WINTER 1878—

Washington, D.C.

THAT A TOTAL SOLAR ECLIPSE WOULD CROSS THE WESTERN regions of North America on July 29, 1878, was inevitable, the natural result of heavenly cycles set in motion eons ago. Eclipses do not occur randomly; they follow patterns known since ancient times. Lunar eclipses can happen only at full moon, solar eclipses at new moon, and both types can take place only within defined "eclipse seasons" that recur every six months or so, shifting slowly backward year by year. A rhythm reigns in the long run as well. Eclipses of a similar character—lunar versus solar, partial versus total—repeat themselves after the passage of precisely 6,585 and one-third days. (This protracted cycle, lasting a shade more than eighteen years, is called the *saros*.) The Babylonians, Greeks, Mayans, and Chinese, among other early civilizations, identified these patterns and used them to predict eclipses, a critical task for a court astronomer for whom the unexpected dimming of the sun or moon might mean the expected loss of his head.

Knowing when an eclipse might occur, however, did not imply the ability to forecast where it could be seen. A lunar eclipse presents little difficulty in this regard—it will always be visible from at least half the earth's surface—and a partial solar eclipse might be seen from a fifth of the planet, but a total solar eclipse is a rare bird indeed. It is visible only within a narrow corridor, called the *path of totality*, which is often little more than a hundred miles wide. Stand within the zone and you will see the moon entirely cover the surface of the sun, converting day to transitory night. Stand outside the zone and you will see a partial solar eclipse—an interesting event, but a fundamentally different experience.

It was not until the eighteenth century that astronomers were able to forecast the path of a total solar eclipse with a modicum of accuracy. The best known of these early eclipse mappers was Edmond Halley, the Englishman who famously predicted the return of a comet that now bears his name. Halley calculated the path of a total eclipse that was about to pass over the British Isles, and he published his map on a broadside titled *A Description of the Passage of the Shadow of the Moon, over England, in the Total Eclipse of the Sun, on the 22 Day of April 1715 in the Morning*. (A rival publisher, apparently more adept at marketing, issued a pamphlet with a catchier title: *The Black-Day, or, A Prospect of Doomsday. Exemplified in the Great and Terrible Eclipse. . . .*) Challenging the rampant superstition of the era, Halley approached the eclipse with scientific intent. He aimed to learn how his prediction compared with reality, so he asked members of the public to report if they experienced darkness at their location and, if so, its duration—"for therby the Situation and dimensions of the Shadow will be nicely determind; and by means therof, we may be enabled to Predict the like Appearances for ye future, to a greater degree of certainty than can be pretended to at present, for want of such Observations." The actual path of the moon's shadow turned out to be slightly off from Halley's prediction—it was wider, angled less northward, and shifted a bit to the southeast—but his map was broadly correct. As predicted, a darkened London fell within the zone.

By the 1800s, eclipse prediction had continued to improve. Astronomers could plot the path of the moon's shadow decades, even centuries, in advance—hence the report in *Nature*, in 1876, alerting the public that an eclipse would visit the western United States on July 29, 1878. Astronomers could calculate an even more exact path as the event drew closer by updating their equations with the latest observations of the position of the moon, whose orbit tended, in the long run, to veer slightly from scientists' expectations. Although to the casual observer the moon's journey around the earth appears as uniform as the movement of a precision timepiece, it is, when examined closely, the walk of a drunkard. The moon weaves and wobbles, jumps ahead and lags behind. Isaac Newton found the moon's motion so puzzling that, as he told his friend Halley, it gave him a headache.

IF ANYONE COULD BE deemed responsible for the total solar eclipse of 1878—tasked with charting its course, calculating its duration, alerting the public—it was Simon Newcomb, the astronomer who had lamented the tenuous state of American science at the time of the nation's centennial. Newcomb had been born in 1835 (a year that, portentously, saw the return of Halley's Comet) not in the United States but in Canada, yet this bookish child from rural Nova Scotia grew up to become an eminent American, one of the most influential scientists of his era. Newcomb was an astronomer, mathematician, physicist, economist, philosopher, professor, government administrator, and popularizer of science all in one, a polymath who exuded power, authority, and a fierce

PROFESSOR SIMON NEWCOMB

intelligence. A student recalled Newcomb as "a big, lusty, joyous man." A friend remembered Newcomb's "piercing eyes, a look full of strength—steady, direct, penetrating." To his sister, Newcomb seemed a force of nature, a man who worked constantly—"like a glacier, slowly, steadily, irresistibly."

Newcomb's primary occupation was as a mathematical astronomer, one who calculated the orbits of the heavenly bodies rather than observed them. Like Isaac Newton, he was obsessed with the motion of the moon, and like Maria Mitchell, he had landed his first paying job in astronomy at the Navy's Nautical Almanac Office, which published the American ephemeris. Whereas Mitchell worked remotely—computing the orbit of Venus from her home in Nantucket and, later, from Vassar—Newcomb worked in the office, then located in Cambridge, Massachusetts. It was less a place of employment than a boys' club. The mathematically inclined young men often set aside their duties to play chess and debate the finer points of religion and philosophy. It seems they also belittled women's intellectual abilities. After Newcomb read a newspaper item about Maria Mitchell—his older and, at the time, more famous colleague—he wrote disparagingly to a friend, "Miss M. is only a female astronomer after all."

Newcomb thrived among this merry band of geniuses and misfits. Although calculating orbits was simple in principle—merely a matter of applying Newton's law of universal gravitation—the task was, in practice, mind-bogglingly difficult. Planets do not blithely revolve around the sun along perfect, elliptical paths; they meander because, as Newton had written, "the orbit of any one planet depends on the combined motion of all the planets." Jupiter pulls on Saturn. Earth tugs on Mars. To derive orbits, Newcomb and his colleagues had to calculate this complex, gravitational dance. "To this work I was especially attracted because its preparation seemed to me to embody the highest intellectual power to which man had ever attained," Newcomb wrote.

Twenty years after beginning his career with the Nautical

Almanac, then leaving for a long stint at the U.S. Naval Observatory, Newcomb triumphantly returned as the man in charge. In
1877, early in the administration of Republican President Rutherford
B. Hayes (whom Democrats mocked as "His Fraudulency" given
the disputed election of 1876, which Hayes had won despite losing
the popular vote), Newcomb was appointed superintendent of the
Nautical Almanac. The office, now in Washington, occupied what
Newcomb called "a rather dilapidated old dwelling-house . . . on
the border line between a slum and the lowest order of respectabil-
ity," but he could not have been happier. He had become the modern equivalent of the ancient court astronomer. It was his job, on
behalf of the U.S. government, to divine the motions of the celestial
spheres, and that included predicting eclipses.

"I WISH TO ISSUE some information respecting the total eclipse of
July 29th next, perhaps a chart," Newcomb wrote in January 1878 to
his assistant George W. Hill, a brilliant computer but socially awkward man. ("Professor Newcomb says Mr. Hill is the finest mathematician in America if not in the world, but he can't say two words,"
a young woman confided to her diary after meeting the timid scientist.) Newcomb instructed Hill "to compute the central line and limits of totality." After plotting the path on a series of maps, Newcomb
published them with accompanying text in a special eclipse bulletin.

"During its progress, the dark shadow of the moon will first strike
the earth in the province of Irkoutsk, Siberia," Newcomb wrote. "Its
course will at first be east-northeast, but will gradually change to
east, and, after leaving Asia, to southeast." The maps showed the
path in detail where it traversed the United States along a gently
curved line. "In this country it will sweep over the western end of
Montana Territory, the Yellowstone National Park, Wyoming Territory, Denver, Colorado, and Northern and Eastern Texas, entering
the Gulf of Mexico between New Orleans and Galveston." Newcomb explained that Americans outside of the path, which was

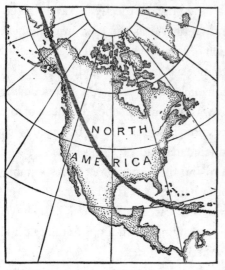

TRACK OF LUNAR SHADOW.

only 116 miles wide, would witness a partial eclipse. He then added this appeal: "Total eclipses of the sun are so rare at any one point that the opportunities they offer for studying the attendant phenomena should be utilized in every way."

What Newcomb meant he explained more fully in *Popular Astronomy*, his new book, which—despite a luke-warm review from Maria Mitchell—sold sensationally, enjoying multiple editions. "Total eclipses of the sun afford very rare and highly prized opportunities for studying the operations going on around that luminary," Newcomb wrote. "[A]s the last ray of sunlight vanishes, a scene of unexampled beauty, grandeur, and impressiveness breaks upon the view. The globe of the moon, black as ink, is seen as if it were hanging in mid-air, surrounded by a crown of soft, silvery light, like that which the old painters used to depict around the heads of saints." Astronomers called this glowing halo the *corona*, and while it is now known to be the sun's extremely hot outer atmosphere, its nature remained a mystery in Newcomb's day. At one time, scientists had thought it might be the moon's atmosphere backlit by the sun, or perhaps it was merely an illusion produced by sunlight passing through the earth's atmosphere. By 1878, the corona was generally accepted as a true solar phenomenon, caused by *something* that surrounded the sun. But what? A massive envelope of luminous gas? A cloud of dust particles that reflected sunlight? A swarm of meteors?

Other odd sights also appeared during a total solar eclipse. "Besides this 'corona,' tongues of rose-colored flame of the most fantastic forms shoot out from various points around the edge of the lunar disk," Newcomb wrote. "They are known by the several names of 'flames,' 'prominences,' and 'protuberances.'" These too had perplexed astronomers.

Ultimately what scientists were trying to unravel was the enigma of the sun itself. What was it made of? What fueled its colossal fires? The phenomena seen at total eclipses provided clues. "We thus have the seeming paradox," Newcomb wrote in 1878, "that most of what we have learned respecting the physical constitution of the sun has been gained by the hiding of it from view."

THE VERY IDEA THAT scientists might discern the physical constitution of the sun and other celestial objects had, until recently, seemed preposterous. "[W]e will never by any means know how to study their chemical composition or mineralogical structure," the French philosopher Auguste Comte, who pondered the limits of scientific knowledge, declared confidently in 1835, when few would have challenged his assertion. After all, mankind could sense the stars and planets only by the light they gave off. How, then, could one possibly take their measure as material entities?

The answer lay in the light itself. Isaac Newton, in the seventeenth century, had famously demonstrated that a prism will break white light into a rainbow. Much later, scientists put this technique to use in the laboratory. They placed various substances into flames and examined the spectrum of light emitted, and found that each chemical element produced characteristic colors. Sodium's light, when passed through a prism, showed two different shades of yellow. Lithium exhibited a brilliant red line and a faint orange one. Potassium glowed in red and violet. Scientists could therefore identify the chemical makeup of any hot, glowing gas by looking for this hidden code in the light it gave off. To conduct their studies, researchers fitted telescopes with prisms, producing spectroscopes. "As the geologist with his hammer breaks up the rocks of earth and determines

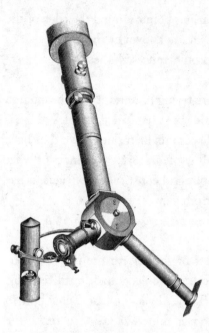

THE SPECTROSCOPE.

their composition, so with the spectroscope the astronomer breaks up the light proceeding from the heavenly bodies, and shows its nature in the colors of the spectrum," explained Maria Mitchell.

The spectroscope offered a new tool to solve the sun's riddles, and by the mid–nineteenth century, astronomers were keen to train it on the mysterious apparitions visible only during a total solar eclipse. Scientists therefore launched expeditions to chase the moon's shadow.

A TOTAL SOLAR ECLIPSE occurs somewhere on earth approximately every eighteen months, but reaching the path of totality is often problematic. Eclipse paths snake across the globe like spaghetti thrown at a map. They traverse oceans more often than land, remote regions more often than populated ones. Some eclipses can be seen only at high latitudes, skirting the North or South Pole.

The scientists who planned expeditions had to examine each eclipse path carefully and consider where to erect observation posts, weighing logistical concerns and climate. An ideal spot must be reachable with a ton or more of scientific equipment, and it should boast good prospects for clear skies, because a single misplaced cloud at the crucial moment could obscure the view and render the entire enterprise worthless. Expeditions required months of planning and weeks of travel, but scientists measured success in mere seconds of observation. In an extreme case, a total solar eclipse may last more than seven minutes, but most last less than three. (The longest eclipses occur when the moon is closest to the earth; at this point in its orbit, the moon appears especially large in the sky and therefore blocks the sun for a greater time.)

In the international race to study solar eclipses, Europeans had taken an early lead. Between 1851 and 1875, scientific parties from England, France, Germany, Italy, and Russia journeyed to Norway, Sweden, Spain, Sicily, India, Ceylon, Siam, the Arabian Peninsula, and Africa (North and South). These expeditions yielded key discoveries. During the eclipse of 1860, British astronomer Warren De La Rue took successive photographs of the rose-colored prominences and showed that they were physical features of the sun, not the moon, which covered them by degrees. In 1868, French astronomer Jules Janssen used his spectroscope to reveal the chemical makeup of those prominences: superheated hydrogen gas. Janssen and others also noticed a puzzling yellow line in the spectrum, which British astronomer J. Norman Lockyer, conducting further study after the eclipse, concluded was due to an as-yet-unknown chemical element in the sun. Lockyer named it helium. (It would take scientists another three decades to discover the element on earth.)

The Europeans returned from their travels not just with discoveries, but with epic tales of derring-do. The British physicist John Tyndall, after observing the Mediterranean eclipse of 1870, recalled braving seas so rough "that my body had become a kind of projectile, which had the ship's side for a target." Norman Lockyer's voyage that same year proved even less enjoyable. His party's steamer, the H.M.S. *Psyche*, struck rocks off Sicily, and although rescuers saved everyone aboard, clouds on eclipse day obscured all but one and a half seconds of totality. But the bravest exploit was that of Jules Janssen, the intrepid Frenchman. Trapped in Paris by the Prussian siege yet determined to witness the eclipse of 1870 in North Africa, Janssen miraculously escaped by balloon, flying over enemy lines at dawn on his way to Algeria.

Americans, too, had chased eclipses with gusto, but with

H.M.S. CALEDONIA AND TWO STEAMERS ATTEMPTING TO TOW
H.M.S. PSYCHE OFF A SUNKEN ROCK ON THE COAST OF SICILY.

considerably less success. The nation's first organized expedition had actually transpired during the tail end of the Revolutionary War. Harvard professor Samuel Williams calculated the path of an eclipse to occur in October 1780, and he found that it crossed the Gulf of Maine in a region held by British forces. John Hancock, the speaker of the Massachusetts House of Representatives and the state's governor-elect, appealed in a gentlemanly manner to the enemy commander "as a Friend of Science" to let the astronomers through. Under the terms of a truce, the British permitted Professor Williams to erect his observation post on an island in the computed path, but—embarrassingly—at the moment when the sun should have been completely obscured, it was not. Either because of Williams's own miscalculations or a faulty map that misstated the island's latitude (Williams blamed the latter), the Harvard expedition had inadvertently placed itself just outside the path of totality.

Eighty years later, a young Simon Newcomb, working at the Nautical Almanac Office, computed the path of an eclipse that would traverse the upper reaches of North America, and he was dispatched to meet the moon's shadow in the wilds of what today is central Manitoba. It was, even by the era's crude standards, an uncomfortable slog: forty-seven days' travel by train, stagecoach, wagon, steamboat, and birch bark canoe, made interminable by squadrons of mosquitoes and a diet of pemmican—dried buffalo meat and rendered fat fashioned into cakes so hardened they required a hatchet for eating. When Newcomb and his companions finally arrived within the belt of totality, they discovered themselves mired in a vast marsh. "The country was practically under water," Newcomb wrote. "We found the most elevated spot we could, took out our instruments, mounted them on boxes or anything else in the shallow puddles of water, and slept in the canoe. Next morning the weather was hopelessly cloudy. We saw the darkness of the eclipse and nothing more." With little to show for their odyssey, the men collected insect, fish, and fossil specimens for Harvard's natural history museum and then reversed the journey, spending the next sixty-three days trudging home.

America's best opportunity to study the hidden sun had come in 1869. That August, a total eclipse crossed obliquely down the eastern half of the country, from Dakota Territory to

OBSERVING THE ECLIPSE.

North Carolina. Congress appropriated $5,000 to underwrite the expenses of astronomers heading to the path of totality, and many did so, including Simon Newcomb, Maria Mitchell, and the rival planet hunters C. H. F. Peters and James Craig Watson, all of whom observed from Iowa. Despite storms the day before, fair weather prevailed for the eclipse, and the astronomers turned their telescopes, spectroscopes, and cameras upon the corona and solar prominences—one of which, appropriately from Iowa, looked like an ear of corn sprouting from the blackened sun. The Americans published a slew of findings, but most were of little note. (One exception was an unexplained green line seen in the spectrum of the corona that some would attribute to an element, like helium, not known on earth. Later dubbed coronium, the substance would eventually turn out to be an unusual, highly ionized form of iron.) British astronomer Norman Lockyer, helium's discoverer, wrote generously and perhaps a bit patronizingly, "[T]he American Government and men of science must be congratulated on the noble example they have shown to us, and the food for future thought and work they have accumulated."

Simon Newcomb heard very different comments through back channels. "[O]ne of our scientific men who returned from a visit abroad [declared] that one of our eclipse reports was the laughing-stock of Europe," he reported.

NOW, IN 1878, another solar eclipse would finally provide a chance for the United States to bolster its scientific reputation. But when European astronomers inquired what the U.S. government planned to do to observe the event, Simon Newcomb could reply only with embarrassment.

"It is doubtful whether we shall have any regularly organized government parties in the field," he wrote to James Ludovic Lindsay, better known as Lord Lindsay, who had organized a private British eclipse expedition to Spain in 1870. "I would be very glad if it could

tempt over some European observers, especially yourself." Newcomb further explained his frustration to Britain's Norman Lockyer.
"[I]t is still uncertain whether Congress will appropriate any money
for the observations," Newcomb noted, adding, "I fear the question
will remain unsettled till it is too late to make efficient preparations."

As an occasional guest at the Hayes White House and a close
acquaintance of the influential Ohio Congressman and soon-to-be
President James A. Garfield, Newcomb likely lobbied in the background, hoping quietly to spur the government to action. The U.S.
Naval Observatory, meanwhile, lobbied openly.

The Naval Observatory, the U.S. government's chief sponsor of
astronomical research, had grown from humble beginnings. In 1842,
Congress had authorized the Navy merely to erect a small facility for
keeping time by tracking the stars, but over the years it had expanded
in size and purview. By the 1870s, the observatory possessed the
largest refracting telescope in the world and housed it in an impressive building, although the
location proved unfortu-
nate given that the nearby
Potomac mudflats plagued
the scientists with fog and
malarial mosquitoes.

As the eclipse of 1878
loomed, the Naval Observatory's superintendent, Rear
Admiral John Rodgers, proposed sending seven gov-
ernment parties to the path
of totality, from Montana
to Texas. Establishing multiple posts, spaced widely,
would increase the odds of
at least one party enjoying
clear skies. Rodgers made

THE GREAT EQUATORIAL—
UNITED STATES NAVAL OBSERVATORY.

his case for funding to the House Committee on Appropriations in March 1878.

"The sun is the source of all light, heat and life existing on the earth," Rodgers wrote. "Many questions relating to the physical constitution of that orb can only be studied during a total eclipse, and as the entire sum of all the opportunities which all the astronomers of the world can get to observe such eclipses does not exceed five or six hours in a century, it seems the part of prudence to utilize these precious moments to the utmost, and not allow one of them to pass unheeded."

Rodgers also appealed to lawmakers' patriotic pride. "Amongst nations, as with individuals in society, one who occupies a prominent place must, if reproach is to be avoided, bear his part in expenses directed to the general well-being. The United States is at once too enlightened, and too important, a member of the confraternity of nations, to avoid a burden which falls naturally to her. If other nations send costly expeditions to the antipodes to observe eclipses, the United States can scarcely refuse to observe one within its own borders."

Rodgers requested money to transport men and equipment to the western frontier. He did not ask Congress to pay salaries; the Naval Observatory would seek astronomers willing to volunteer for duty. The total request: $8,000, a pittance by modern standards, hardly much more in 1878.

The House declined to fund it.

"This seems to be another instance of unjust discrimination against the far West," railed *The Denver Daily Tribune*. "If the approaching total eclipse had been appointed for the White Mountains, for Boston, for New Orleans, for the peaks of the Virginia mountains, for Philadelphia, for Chicago, San Francisco, Kansas City, or even perhaps Topeka, Congress would have treated it with proper respect and due consideration, would have recognized its claims, and would have made ample provision for its observation. In fact, Congressmen would no more have allowed it to go unobserved than they do the handsome female lobbyists that visit their hall."

The *New-York Tribune* was inclined to agree, similarly rebuking Congress for its inaction. "The lame excuse for this is that the committee were not fully informed as to the importance of the matter; let us hope that, when it is again brought before Congress, there will be no chance for the plea of ignorance on the part of our legislators."

Debate moved to the Senate, where lawmakers discussed adding the funding to a House appropriations bill. Maine Senator James G. Blaine, who a few years later would serve as Secretary of State under President Garfield, favored the amendment on geopolitical grounds. "[W]e should feel in a very awkward condition to have the French Academy and the Royal Observatory at London and other scientific societies sending their agents over here for the eclipse to be observed within our own limits, and our own Government taking no notice of it whatever." Minnesota Senator William Windom, who would become Garfield's Secretary of the Treasury, opposed the amendment for fiscal reasons. He argued there were higher priorities for federal funding, most notably the District of Columbia's impoverished schools.

"[Senator Blaine] proposes to send a commission off skylarking at an expense of $8,000 to watch the eclipse of the sun while he is eclipsing these young intellects right here under the Dome of the Capitol," Windom declared. "There is too much 'moonshine' in this."

As the winter of 1878 yielded to spring, as the heavenly bodies moved toward their midsummer appointment in the western skies, the country remained unready—not only for the eclipse, it seemed, but to fulfill its responsibility as an enlightened member of the global scientific community.

CHAPTER 6

THE WIZARD
IN WASHINGTON

THURSDAY, APRIL 18, 1878—
Washington, D.C.

LEAVES WERE BUDDING, AND SOON ALL WAS GREEN ALONG the National Mall, the patchwork of parks that stretched west from the U.S. Capitol to the Potomac. The Washington Monument, three decades after its cornerstone had been laid, remained a mere stump due to a chronic lack of funds for its construction, but the Smithsonian Institution stood complete and ornate thanks to Old World largesse. The red sandstone castle, clad in ivy and embellished with towers and battlements, had been built from the bequest of an Englishman who aimed to promote "the increase and diffusion of knowledge" in America. Inside, it overflowed with fossils and minerals, live snakes and Peruvian mummies, and innumerable bird carcasses and mammal skins that infested the building with fleas.

In this third week of April, in a first-floor room devoted to geographical reports, a "noon repast" was served. The centerpiece: a thirty-pound boiled salmon garnished with vegetables that had been carved like flowers (turnip japonicas and carrot marigolds). Then the lectures resumed. Simon Newcomb outlined plans for an elaborate experiment to measure the velocity of light. Others spoke of recently discovered fish and reptile fossils, and of new theories to explain the workings of atoms and molecules. The spring meeting of America's National Academy of Sciences was in session.

The academy was America's most elite—and elitist—assemblage of scientists, a congressionally chartered body of experts self-consciously modeled after Britain's Royal Society and the French Academy of Sciences. Created just fif-

MAIN ENTRANCE, NORTH FRONT, SMITHSONIAN INSTITUTION.

teen years earlier, ostensibly to advise the government but as yet rarely called upon to do so, the academy was still trying to find its bearings. It did little more than hold semiannual meetings and publish memoirs of its deceased members. Simon Newcomb called it "about the sleepiest and slowest institution with which I am acquainted."

The individual members, however, were among the most influential and accomplished scientists in the nation. They included those dueling asteroid hunters C. H. F. Peters and James Craig Watson; geologist Ferdinand V. Hayden, whose 1871 survey of Yellowstone had prompted the creation of America's first national park; paleontologist Othniel C. Marsh, a Yale professor whose excavations in the American West were unearthing bizarrely giant fossils ("The name of the animal," one of his assistants explained to a journalist, "is what is termed the Dinosaur"); and Joseph Henry, secretary of the

Smithsonian, who, though aged and frail (he would die a month later), served as the academy's president. Conspicuously absent from the rolls was Maria Mitchell—the academy included no female members and would not elect its first until 1925—yet, in the very room where the distinguished men assembled, her portrait stared down from the wall, presumably put there by Joseph Henry, who had long been an ally and mentor.

THE LATE
PROFESSOR JOSEPH HENRY.

Physicist George F. Barker, Thomas Edison's friend from the University of Pennsylvania, was a member of the academy, and during his visit to Washington for the meeting, Barker lodged at Simon Newcomb's home, a three-story brick row house decorated in mahogany and furnished with a guest room on the top floor, tucked in the back. Funding for the upcoming eclipse was still in doubt, and it was a topic the men surely discussed. "I shall be only too glad to do anything I can to help on your appropriation for the Eclipse expedition," Barker had written to Newcomb the previous week. "How can I be of service?" When not engaged in scientific conversation in the parlor or dining room, Barker joined Newcomb's young daughters in the hall for a game of blind man's buff. The girls were fond of Barker's shaggy sideburns. They liked to braid his whiskers beneath his chin.

On this, the third day of the meeting, after the lunch and lectures, the academy suspended its regular session for a special presentation at ten minutes past four in the afternoon. The scientists, then joined by wives, children, and guests, crowded into Joseph Henry's office, and the audience grew so large that the doors had to be lifted from their hinges to allow spectators to watch from the hallway. At the front of the room, clearly uncomfortable with all the eyes cast upon him, sat Thomas Edison, twisting a rubber band between his fingers. The scientific men, sporting generous beards, appeared wise and scholarly. Edison—clean-shaven and with tousled hair that, as one observer noted, "stood out at all angles in defiance of comb rule"—seemed of another generation. Indeed, of a different era.

The object of all the attention was not the man, but his machine, which consisted of a tinfoil-covered brass cylinder that was set horizontally and rotated on a shaft beneath a small mouthpiece. Edison's assistant placed the contraption on a desk and turned the crank. The device emitted a faint but clear human voice. "The speaking phonograph has the honor of presenting itself before the Academy of Sciences," it said in a tone both nasal and seemingly distant. Children pressed nervously against their parents, while adults stared at each other in awe. "I declare," remarked a member of the audience who had apparently heard a demonstration before, "it sounds more like the devil every time."

THE PHONOGRAPH—AN INVENTION both astonishingly simple and, simply, astonishing—had propelled Thomas Edison from relative obscurity to international celebrity in the early months of 1878. New York journalists descended on Edison's country workshop to write gushing profiles. "The Napoleon of Science," they called him. "The Jersey Columbus." "The Wizard of Menlo Park." (And one journalist now referred to Menlo Park as "Edisonia.") Newspapers nationwide picked up the copy and ran with it. "The Mania has broken out this way," one of Edison's business associates penned from Chicago. "School-girls write compositions on Edison: The funny papers publish squibs on Edison: The religious papers write editorials on Edison."

When Edison first announced his invention of the phonograph,

THE PHONOGRAPH AND ITS INVENTOR, MR. THOMAS A. EDISON.

several months earlier, many had insisted it was nothing more than a ventriloquist's trick. (A stalwart college professor wrote, "The idea of a talking machine is ridiculous.") The notion that one could capture sound and release it at will seemed magical, like dabbling in the dark arts. After all, the phonograph was mere metal—no lips, no teeth, no tongue—and yet it spoke, enabling a person's voice to live beyond the grave. "Speech has become, as it were, immortal," commented *Scientific American.* Everyone wanted to see the device, to hear it, to talk about it.

Earlier in the day at the Smithsonian, a Washington newsman had approached Edison. "Let me, like all the rest of the world, congratulate you on your discovery of the phonograph," the reporter said as Edison blushed and bowed his head. "What a pity you hadn't invented it before. There is many a mother mourning her dead boy or girl who would give the world could she hear their living voices again, a miracle your phonograph makes possible."

"No, I don't think the world was ripe for it before," Edison said.

"Were you a long time in perfecting the discovery?" the writer inquired.

"Oh, no; I had thought of the idea vaguely many times, long before I undertook to work it out," he said. "It is a very simple idea when you come to look at it, and the wonder is it wasn't discovered before."

"How old are you, Mr. Edison?" the reporter asked.

"Thirty-one."

"Very young yet."

"I am good for fifty," Edison brazenly replied, "and I hope to astonish the world yet with things more wonderful than this."

DURING HIS EXULTANT VISIT to Washington, Edison demonstrated his phonograph to congressmen, senators, and, during a late night drop-in at the White House, to President Hayes and the cultured, college-educated First Lady Lucy Webb Hayes. Yet Edison's scientific audience, at the Smithsonian, was arguably the most important.

Yale professor O. C. Marsh, the dinosaur specialist, formally welcomed Edison on behalf of the academy. Edison, who was partially deaf—a surprising infirmity for the man who had just invented the phonograph—held his right hand to his ear, yet he still could not make out what was said. Given his hearing impairment and aversion to crowds, Edison asked his friend George Barker to speak on his behalf. Barker told his colleagues that Edison was "a man of deeds, not words." Edison's assistant continued the demonstration— singing, whistling, and crowing into the device, then playing the snippets back as if thawing frozen echoes. Sheets of tinfoil, etched with the dents and pinpricks that recorded the sound, were passed around for inspection. The great men applauded. Edison rose to take a sheepish bow.

Once the phonograph performance had ended, Barker introduced another of Edison's recent inventions—a new kind of telephone that threatened to upstage Alexander Graham Bell's device. Barker recalled the stifling day at the Philadelphia Centennial two years earlier, in June 1876, when Bell had amazed the judges with his telephone. In fact, Barker said, it had so excited Britain's Sir William Thomson that he shouted himself hoarse through it. But this revealed a fundamental flaw with Bell's telephone; one *had* to shout through it because the mouthpiece produced a weak signal that was barely audible at the other end. Edison had now devised a solution.

At his workshop in New Jersey, Edison discovered an intriguing property of carbon: the element's electrical resistance varies dramatically with pressure. Edison then put this discovery to use. He collected the carbon soot produced from burning kerosene and other lamp oils, and he pressed it into small disks that he called carbon "buttons." These he placed in the mouthpiece of his telephone. When connected to a battery, the carbon turned the vibrations of the human voice into an electrical signal, one strong enough to elicit clear, loud sound from the telephone receiver.

George Barker presented his colleagues with an example of Edison's carbon telephone. It was connected by wire from Washington to a telegraph office in Philadelphia, and a conversation was initiated between the cities 135 miles apart. A voice down the line called out, "Halloo, halloo." (The use of this word as the standard telephone greeting, which evolved into today's "hello," was Edison's idea. Alexander Graham Bell had advocated a different word: "ahoy.") The scientists took turns testing Edison's telephone and comparing it with others designed by Bell and rival engineers that were also on display. Barker praised the superiority of Edison's device, and of Edison's intellect. "You do well, Mr. Edison, to claim you have discovered the principles of science upon which your telephone is based. These other men are only inventors," Barker said. "You, sir, are a discoverer!" The audience applauded robustly.

EDISON NO DOUBT APPRECIATED the praise from America's scientific elite, as he still endured ridicule for his experiments on the etheric force. *The New York Times* had recently seen fit to remind its readers that the man who invented the phonograph "not very long ago . . . became notorious for having discovered a new force, though he has since kept it carefully concealed, either upon his person or elsewhere." Edison seemed more intent than ever on proving himself a real scientist, worthy of respect.

Earlier that week, a celebrated journalist had spent two hours at Menlo Park interviewing Edison about the telephone, the phonograph, and the progress of science. "I think that science is the greatest interest of the present and future: of more national value than armies or navies," Edison said. And by "science" he meant more than the engineering of practical gadgets.

The journalist asked Edison, "What position does Morse take in your mind?" referring to Samuel Morse, the American painter-turned-entrepreneur whose electrical device and communications code had changed the world. "I suppose he holds his rank as the inventor of the telegraph."

Edison's response was revealing.

"No, he was hardly an inventor. He belongs to the class of 'promoters,'" Edison sniffed. "The scientific man was Professor Joseph Henry, now of the Smithsonian Institute. He made the discoveries which produced the telegraph." Indeed, Henry's work with electromagnets had provided key insights that Morse exploited to build, patent, and profit from the telegraph. Edison expressed frustration that Henry, who toiled selflessly to unlock nature's secrets, had not received more credit or fame. "The newspapers mixed it up," Edison said, "and got the reputation over to Morse."

Edison seemed determined to show that he was more Joseph Henry than Samuel Morse—a discoverer, not a crass promoter—in both his public statements and his personal relations with other scientists. He had recently contacted none other than Charles Darwin with an offer to help the Englishman's studies of natural selection. Edison had noticed a peculiar-smelling bug at his Menlo Park laboratory. "I suppose this odor is used as a means of defence [sic] like that from the skunk," he wrote. "I could procure some [insects] next summer and send by mail." Darwin duly replied with his thanks but declined the offer.

Edison found better luck engaging with astronomers. One of them, Samuel Pierpont Langley, directed the Allegheny Observatory in Allegheny, Pennsylvania, just across the Allegheny River from Pittsburgh (and now part of it). Langley specialized in research on the sun, due in part to necessity—the industrial city's noxious air pollution all but obscured the stars—but also as a result of his

PROFESSOR S. P. LANGLEY

boyhood fascination. "I used to hold my hands up to [the sun] and wonder how the rays made them warm, and where the heat came from and how," he recalled. This query became the focus of his career. Langley studied the sun's radiant heat. He examined how much energy came from different areas of the solar surface and was carried by different colors of the spectrum, including invisible infrared rays. For his research, Langley used a sensitive electrical thermometer called a thermopile, but as his studies grew more sophisticated, he needed a measuring device that was even more responsive. So he turned to Edison for help. "If you could make something . . . say one hundred times as sensitive," Langley wrote, "you would not perhaps produce anything commercially paying, but you certainly would confer a precious gift on science."

Edison took up the challenge, and toward the end of the day at the Smithsonian, after the demonstrations of the phonograph and the carbon telephone, he told a Washington newsman of this new focus of his inventive energy. The remark came in passing, but it signaled a consequential shift in Edison's life and career, one that would involve the Wizard of Menlo Park in the eclipse of 1878.

"Have you made any recent improvements or discovered any new applications of your instruments?" the journalist asked.

"Well, nothing recently, so far as the phonograph is concerned," Edison said. "Night before last I found out some additional points about the carbon which I use in my carbon telephone. It may be used as a heat measurer. It will detect one fifty-thousandth of a degree Fahrenheit." Edison's eyes brightened and his face grew animated as he discussed the use for such an instrument in astronomy.

"I don't know but what I can make an arrangement by which the heat of the stars will close the circuit at the proper time automatically and directly. It is a curious idea that the heat of a star millions of miles away should close a circuit on this miserable little earth, isn't it?" Edison said. "But I do not think that it is impossible."

CHAPTER 7

SIC TRANSIT

MONDAY, MAY 6, 1878—

Ann Arbor, Michigan

JAMES CRAIG WATSON DID NOT ATTEND THE SPRING MEETING of the National Academy of Sciences in Washington. He was busy in Ann Arbor, where life proved gratifying for the corpulent narcissist. "There has been an unusual interest taken in Astronomy this year, and another whole class are sworn admirers of Prof. Watson and his mode of teaching," the University of Michigan student newspaper proclaimed. Outside the classroom, at the observatory, Watson enjoyed a rapid string of successes; he had recently discovered three new asteroids and now ranked second in the world in the planet-hunting race, tantalizingly just behind his rival, C. H. F. Peters. And Watson's skill had been acknowledged by Washington when he received an official request to help observe an important astronomical event in early May. "[T]he Naval Observatory desires, if possible, to secure your cooperation in the work," it read. That "work" involved research on a pair of planets, one real, the other hypothetical.

The irrefutably real planet was Mercury. As the innermost known world in the solar system, Mercury occasionally passes directly between the earth and sun, a conjunction similar to a solar eclipse. Unlike the moon, however, Mercury appears far too small in the sky to reduce the amount of sunlight by any noticeable degree. What one sees, if one knows to look, is the silhouette of a planet—a perfect black freckle that slowly traverses the bright face of the sun. Astronomers call such an event a *transit*.

Transits of Mercury occur infrequently, just thirteen or fourteen times a century, while those of Venus are even rarer. Because of the alignment of orbits, Mercury can cross the face of the sun only in November or May.

On the first Saturday in May of 1878, the *New-York Tribune* alerted readers that just such an event would occur in a couple of days:

Next Monday the astronomers of Europe and this country will bend their energies and point their telescopes toward the planet most nearly bathed in solar fires. Perhaps, however, after this investigation, such a phrase will no longer describe Mercury. For the main object to be attained by Monday's research, is to determine whether or no there is some other planet or planets nearer the sun.

For two decades, this question had cleaved the astronomical community: Does another planet circle the sun within the orbit of Mercury? Some scientists were convinced that such an *intra-Mercurial* planet did exist. They gave it a name—Vulcan, for the Roman god of fire—and listed it in encyclopedias and textbooks as a legitimate member of the solar system's family. Such pronouncements sparked public imagination about life on a world where, given the planet's hurried orbit, the years raced by. "The Fourth of July must return with maddening rapidity, and the Vulcanites must be scourged with Centennial Exhibitions at least four times as often as the inhabitants of any part of our slower and more considerate planet," mused *The New York Times*.

Another camp of astronomers, though, "declined to believe in the existence of this planet," the newspaper explained. The skeptics scoffed at the believers as if they were "pretended discoverers of the sea-serpent. In fact, the alleged planet Vulcan was looked upon very much in the light of an astronomical sea-serpent."

No astronomer was more skeptical than C. H. F. Peters, the world's most prolific asteroid hunter. "Professor Peters of Hamilton college, the indefatigable discoverer of new stars . . . pronounces [Vulcan] a myth," one newspaper reported. "While setting out his own views . . . he takes pains to say that Professor Newcomb of Washington, at one time swallowed this intra-Mercurial planet whole, so to speak, but it is a bait he has himself always scorned."

Newcomb, in reality, had long expressed doubts about Vulcan's existence, but he kept an open mind. James Craig Watson, on the other hand, had firmly taken the bait—and the proverbial hook, as well. "Prof. Watson has long been a believer in the existence of a planet inside the orbit of [M]ercury, as those attending his lectures in past years are fully aware," the University of Michigan's student newspaper explained.

The U.S. Naval Observatory hoped that the 1878 transit of Mercury, carefully observed, would resolve this dispute. Instead, the day's events would only inflame it.

THE IDEA THAT VULCAN existed had arisen from earlier observations of Mercury, a planet whose orbit is swift and eccentric, tracing a path more oval than round. Mercury's closest approach to the sun, a point called its *perihelion*, shifts incrementally forward with each passage, so that as the planet wheels around the sun, its orbit makes its own gradual orbit.

That Mercury's perihelion would advance slowly could be explained by the gravitational attraction of the other known planets, but the magnitude of this effect did not match scientists' calculations. "Mercury has, or has been thought to have, defied the ordinary rules for a planet's behavior," wrote a Chicago newspaper. "[T]he point of perihelion in the orbit of Mercury has moved from west to east more rapidly than is allowed by the figures of the mathematicians."

The mathematician who had studied this mystery most carefully was the late director of the Paris Observatory, Urbain Le Verrier (sometimes Anglicized "Leverrier"), the same man who had snubbed Maria Mitchell on her visit to France. A brilliant yet imperious scientist—"Even members of the [French] Academy could not suppress their detestation of him," Simon Newcomb recalled—Le Verrier had established his genius in 1846 when he discovered the planet Neptune, not with a telescope but by sheer calculation. (As a colleague put it, he found the planet *au bout de sa plume*, at the tip of his pen.) Le Verrier had accomplished this feat by studying the motion of another world—Uranus. Previously the outermost known planet, Uranus was found to veer perplexingly from its predicted orbit, and Le Verrier concluded that the strange behavior could be explained by the gravitational attraction of an unknown planet residing even farther from the sun. He then calculated the size and location of this hypothetical object based on its effects. When astronomers in Berlin aimed their telescope at the appointed corner of the sky, they found Neptune. It was a triumph of Newtonian mechanics and French skill.

Later, Le Verrier turned his attention to Mercury. As he had done with Uranus, he calculated what kind of object would explain the perturbations in Mercury's orbit, and he prophesied the existence of Vulcan (or, perhaps, multiple Vulcans). In 1859, he called on astronomers to search for such an intra-Mercurial planet, but finding a world at the extreme inner reaches of the solar system presented problems. Sitting so close to the solar disk, Vulcan must rise and set with the sun, and therefore it would not be visible at night—perhaps not even at dawn or dusk. Like Mercury and Venus, however, Vulcan should be seen, on occasion, in transit.

Over the years, some astronomers claimed to see just that—an unknown planet sailing across the solar face—but these supposed transits of Vulcan were always suspect, reported by amateur astronomers using inferior equipment. In several cases, the observers had likely seen sunspots that they mistook for the hypothetical planet. In no cases were the observations corroborated.

"[D]uring the last ten or fifteen years the sun has been studied so assiduously by professional astronomers that they necessarily would have fallen in with a transit if a planet at a distance from the sun less than Mercury's existed," C. H. F. Peters concluded. "We have to consider, therefore, the non-existence of such a planet or group of planets as a question set at rest." As Peters saw it, Vulcan had been called into existence for one reason—to solve a small mathematical discrepancy in Mercury's orbit—and a more logical solution to the mystery lay not in inventing a new planet but in questioning whether there was a problem that needed solving at all. Given Mercury's proximity to the sun, the planet was difficult to observe (so difficult, in fact, that Copernicus was said never to have seen it), and its orbit was therefore hard to pin down. Perhaps historical calculations of Mercury's orbit had been inaccurate. Perhaps new, more exact observations would show that Mercury's motion was perfectly explainable without the need to conjure a new planet.

Such was the importance of the upcoming transit of Mercury— it now offered an opportunity to measure the planet's orbit with unprecedented precision. As Simon Newcomb explained, the event should answer "whether the result of LE VERRIER, that the motion of the perihelion of Mercury is much greater than that due to the action of the known planets, is really correct." If new observations showed Le Verrier to be wrong, then Vulcan would vanish as it had first appeared—in a mathematical puff of smoke.

EARLY ON MONDAY, MAY 6, the sun lifted into a clear sky over the fields and treetops of eastern Michigan. James Craig Watson had spent weeks preparing for this day. Behind the Detroit Observatory he had erected a photoheliograph—essentially a forty-foot-long horizontal telescope that, by means of a rotating mirror, projected a stationary image of the sun onto a photographic plate. With this, he and two assistants planned to take dozens of pictures of Mercury's slow traverse, which would last seven and one half hours. First, however, they had to identify the precise time when the event began.

During any transit or eclipse, astronomers are careful to note several key junctures, called *contacts*. First contact would be the moment when Mercury's outer edge—its *limb*—first appeared to touch the limb of the sun. (Maria Mitchell described first contact as "the exact instant when an unseen spherical body appears to touch a seen spherical body.") At that time the planet, viewed through a telescope, would look like a tiny notch on the sun's external edge. Second contact would occur a few minutes later when the entire sphere of Mercury could finally be seen, like a bead of black ink about to drip from the sun's edge into the fiery interior. Third and fourth contacts would come at the end of the transit when, respectively, Mercury reached the other limb of the sun and then exited the solar disk altogether.

If Le Verrier's computation of Mercury's orbit was correct, first contact should occur just shy of 9:38 A.M. Ann Arbor time. Watson opened the observatory dome and swung the large equatorial telescope into position, aiming it at the sun's eastern limb. He focused his gaze through an eyepiece that magnified the heavens by a factor of 400 (the light having already passed, presumably, through a dark filter to protect his vision). He watched for that tiny dent—evidence that Mercury had begun its ingress. Accurate timing was essential. The seconds ticked by.

Across the country, other eyes and telescopes were similarly fixed on the sun. In Poughkeepsie, Maria Mitchell used the transit as a teaching opportunity. She assembled her astronomy class at the Vassar observatory to view and photograph Mercury. Farther upstate, at Hamilton College, C. H. F. Peters entertained a broader public. He cast an image of the sun onto a paper screen and narrated the unfolding drama to assembled townsfolk, students, and faculty. When a small black dot appeared on the sun's disk, Peters pointed.

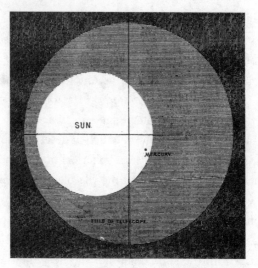

FAC-SIMILE OF A PHOTOGRAPH OF THE TRANSIT.

"There, professor," he said to Edward North, who taught Greek, "you see now what Copernicus never saw." In New Jersey, Thomas Edison—newly passionate about astronomy—had borrowed a telescope to view the event from his Menlo Park laboratory.

Edison's friend George Barker joined a team of scientists at a private observatory owned by Henry Draper, a New York University professor whose telescopes inhabited a twin-domed building north of the city, on the east bank of the Hudson River. Barker, like James Craig Watson, strove to time precisely when Mercury first appeared on the sun, but the observing conditions were far from ideal. Although the skies were clear, instability in the atmosphere caused telescopic images to jiggle as if viewed through vapor rising off a boiling pot. Mercury danced so erratically that Barker was unable to pinpoint the moment of first or second contact. He expressed his frustration on this otherwise-beautiful spring day with a Latin-inspired pun. "Sick transit," he said, "glorious Monday."

In Ann Arbor, however, Watson enjoyed an unhindered view of

DR. DRAPER'S OBSERVATORY AT HASTINGS-ON-THE-HUDSON.

Mercury. "[T]he planet appeared as a sharply defined black disk on the sun," he wrote, and the precisely timed contacts seemed to leave little doubt as to the day's conclusion. "The fact that the planet made its first contact with the sun within nine seconds of the computed time shows that the calculations, upon which Leverrier based his theory, were accurate," wrote *The Detroit Free Press.* "Therefore the unexplained motion of the perihelion of Mercury is not due to inaccurate observations and calculations, but to an unknown planet." The newspaper then added: "Prof. Watson thinks that the result of Monday's observation cannot fail to convert even Dr. Peters to Leverrier's theory."

Such was not the case—"Dr. Peters reiterates his disbelief that there is such a planet as Vulcan," read a widely printed article—but many other scientists moved toward Watson's camp. "As the truth of Leverrier's discovery of an apparently unexplained motion of the perihelion of Mercury is now established beyond all doubt," the U.S. Naval Observatory announced, "it is important to renew the search for an intra-Mercurial planet or planets."

One of the few opportunities to search for Vulcan, a planet so close to the sun that it could never be seen at night, would occur in just a few months' time—when the moon hid the sun from view, during the brief midday darkness of the solar eclipse.

"DEAR SIR: I HAVE just heard that we have a sum of money appropriated for the Solar Eclipse of July 29th," read the timely letter from the superintendent of the U.S. Naval Observatory, Rear Admiral John Rodgers. Congress had finally approved that munificent sum of $8,000 to send expeditions to the West, and the government now sought volunteers. Rodgers's form letter, copied with Edison's electric pen, went to astronomers at America's top universities and observatories. "Please inform me whether you are inclined to assist in making the desired observations, and which ones you would prefer to make. If you accept, then Railroad tickets to and from your destination, and five dollars per day, will be allowed for expenses." The replies, marked "accepts" or "regrets," came in like responses to a wedding invitation.

"It is a great temptation for me to go West for the Eclipse, but I ought not to go," answered C. H. F. Peters, who was tied up with university obligations and was clearly uninterested in standing in the moon's shadow to look for a planet he was convinced did not exist. Furthermore, newspapers reported growing tension in Idaho's Snake River country between white settlers and the Shoshone-Bannocks, the result of broken treaty promises and inadequate government food rations that left the tribes hungry. "So, you go to Montana," Peters wrote to a friend at the Naval Observatory. "Take care of not being scalped by the Indians!" The Bannock War soon erupted, and the Naval Observatory scrapped its planned expeditions to western Montana.

Despite the unrest, James Craig Watson eagerly agreed to journey to the frontier. "I accept your polite invitation to take part in the Observations of the total eclipse," he wrote to Admiral Rogers, making it clear that Vulcan was his quarry. "I think I had better search for the intra mercurial planet during the totality, as so many observers will pay attention to the Corona."

Other responses flowed in as well, from Harvard, Yale, Princeton, Johns Hopkins, the University of Rochester, and observatories elsewhere. Noticeably missing from the list of invitees, however, was the observatory that Maria Mitchell oversaw at Vassar, that brazenly upstart female college in Poughkeepsie.

CHAPTER 8

"GOOD WOMAN THAT SHE ARE"

———

SATURDAY, JUNE 22, 1878—

Poughkeepsie, New York

R AISED ON AN ISLAND, MARIA MITCHELL WAS PERHAPS USED to being an outsider, but she never accepted her exclusion from science's inner sanctum, a sex-based segregation that continued even as her fame grew. This discrimination was often promulgated by sympathetic men who professed respect yet were blind to their own condescension.

In April 1878, Mitchell had visited the Smithsonian Institution, although not, of course, for the meeting of the all-male National Academy of Sciences. Rather, she had met the prior week with leaders of the Woman's Congress to plan for that group's annual convention, to be held in Providence in October. Before going to Washington, Mitchell had reached out to her supporter Joseph Henry, head of the Smithsonian, seeking a space for her humble gathering of prominent women. "Our meeting will be for a few hours only," she wrote. "Even, for so small a meeting, it would help us in the eyes of many who do not investigate, if we were under the roof of a building devoted to the diffusion of knowledge."

Henry responded awkwardly. He consented to the use of a

Smithsonian office, but on the stipulation that Mitchell "not . . . publish an account of your meetings in the newspapers." His stated excuse was practical; he did not want to encourage a flood of requests from other organizations seeking meeting space. The subtext was implied. He likely did not want to be seen supporting a group that so flagrantly promoted women's rights and abilities.

Mitchell's sex—and that of her students—had been used, in a more bald-faced way, to block their participation in a celebrated scientific undertaking just a few years earlier. Astronomers long knew that December 1874 would bring an exceedingly rare occurrence, a transit of Venus, the first in more than a century. By carefully observing the event from different parts of the globe, scientists hoped to calculate a fundamental unit of the solar system—the distance from the earth to the sun. The transit would not be visible from North America or Western Europe, however, so the United States made plans to send astronomers to Asia, the South Pacific, and the Indian Ocean. (Russia, England, France, and Germany made similar plans.) Among those chosen to lead expeditions were the inveterate planet hunters James Craig Watson, who would travel to China by commercial steamship, and C. H. F. Peters, whose destination was New Zealand via Navy vessel. The expeditions sought young scientists as assistants, and in early 1874 Maria Mitchell inquired if her students might participate. Mitchell's old boss and friend Charles Henry Davis, founding superintendent of the Nautical Almanac Office and now head of the U.S. Naval Observatory, replied, "[I]t would be absolutely out of the question to make a cultivated woman decently comfortable on board of a man-of-war." He also ruled out sending women by commercial transport. As he asserted with Victorian bombast, "I should hardly think it worth while to expose a woman to the fatigue, hardships and dangers of so long a winter's journey, unless her services were actually indispensable." The short answer was no. In the end, one woman was permitted to accompany the expedition to China, but not in an official capacity, that being James Craig Watson's dutiful wife.

Now, in 1878, Mitchell found herself yet again excluded. The U.S. Naval Observatory, having received its appropriation from Congress, rapidly scaled up plans for eclipse expeditions and eased the way for astronomers to participate. It bought them train tickets, loaned them telescopes and other instruments, and arranged to ship the equipment—weighing an estimated 10,000 to 15,000 pounds—ahead of time in a specially guarded freight car. (It was a Pennsylvania Railroad fast mail car, "fitted up in a style suited to the requirements of a post office," with boxes and bins for sorting letters.) The U.S. government even extended courtesies to foreign astronomers, ensuring they could bring their scientific equipment into the country duty free and encouraging railroads to sell train tickets at a discount to scientists from abroad.

Maria Mitchell wondered if any of these benefits might naturally extend to her. Undeterred, she wrote to Admiral Rodgers, the new head of the Naval Observatory. "Will you please inform me if the reduction in railroad fares for the benefit of foreigners who go out to Denver to observe the Eclipse, will be granted to American scientists?" Rodgers received a similar appeal from an amateur astronomer in Fort Dodge, Iowa—an enthusiastic fellow named Fred Hess, who had been inspired by Simon Newcomb's *Popular Astronomy*—and Rodgers obliged him by forwarding an official letter that Hess could show to ticket agents in the hopes of getting free or discounted transportation. Rodgers responded less helpfully, however, to Mitchell, despite her being a professional astronomer at a college. "It is very certain that if any one should have the facility offered, of going to observe the Eclipse, at a reduced cost, you are prominently among the number," he wrote, yet he suggested there was nothing he could do to assist. "I would advise you to write to the President of some one of the through lines; and I hope he will be as reasonable and as appreciative to you as the Penn. R.R. has been to foreigners."

Mitchell was increasingly convinced that if women were to succeed in science and in higher education, they could not rely on men but had to join together to help one another. "It is time that women worked in earnest for the education of women," she had recently written. As a professor at Vassar, she saw it as her role not only to teach female students but to foster a sense of community, to create the kind of supportive environment for intelligent women so lacking in the outside world.

THE SCHOOL YEAR WAS ENDING, and Vassar was preparing for its twelfth commencement. The coming days would be filled with obligatory prayers and processions, concerts and speeches. "The college is handsomely decorated," a Poughkeepsie newspaper reported, "and every necessary preparation has been made to give visitors a warm welcome and the graduating class of '78 a grand send-off."

On this day, however—Saturday, June 22—the weather was less welcoming, the damp skies turning the green campus gray. There was little reason to go outside—most college life took place in a massive, five-story building that housed dormitories, classrooms, a dining hall, chapel, and library—but at seven-thirty in the evening, during a pause in the rain, some of the young women ventured out. They followed a broad path toward the northeast, where the college observatory, graced with a lamppost and cast-iron staircase out front, sat atop a knoll. Professor Mitchell was there, awaiting her guests, who had been specially invited.

Mitchell ushered her students, mostly juniors and seniors, inside and into the dome. The circular room normally appeared spare and workmanlike. The curved ceiling formed an unfinished rotunda, exposed ribs of pine supporting a tin roof. In the center of the room, on a massive stone pier, stood the equatorial telescope with its complex assemblage of gears and pulleys. But on this evening, beneath it all, the space had been transformed. Tea tables surrounded the great astronomical instrument. Place cards and roses marked each seat. Candlelight flickered in the hemispherical void.

The occasion, a highlight of many a Vassar student's college career, was Professor Mitchell's annual "dome party." The evening

THE OBSERVATORY.

began with refreshments—usually ice cream, cake, and summer fruit. ("I shall always believe that strawberries have the flavor of a new variety, when eaten with surroundings so unique," a student later reminisced.) Then came the poems. Mitchell passed around a basket containing rolled slips of paper onto which she had written short, silly, but also unmistakably political rhymes. Each praised one of her students. (An example from a previous year: "Here's to our Jessie/ So gay and dressy/ So full of capers and pranks,/ Yet a woman of views/ She would not refuse/ To vote, if she had a chance.") The young women selected verses at random and read them aloud, to the delight of all.

The students then turned to praise their illustrious professor. They sang a tune modeled after Julia Ward Howe's "Battle Hymn of the Republic," which in turn borrowed its melody from the Civil War anthem "John Brown's Body":

> *We are singing for the glo-ry of Ma-ri-a Mitchell's name,*
> *She lives at Vassar College, and you all do know the same.*
> *She once did spy a com-et, and she thus was known to fame,*
> *Good woman that she was.*

Glo-ry, glo-ry, hal-le-lu-jah!
Glo-ry, glo-ry, glo-ry, hal-le-lu-jah!
Glory, glo-ry, glo-ry, glo-ry, hal-le-lu-jah!
Good woman that she was.

The song, written by Vassar students a few years earlier, had become a dome party tradition. As the verses continued, they descended into folly. "[T]he English language was ransacked and grammar set at defiance (in the interests of rhyme)," a guest once noted.

She leads us thro' the maz-es of hard As-tron-o-my,
She teaches us Nu-ta-tion and the laws of Kep-ler three,
Th' inclination of their or-bits and their ec-cen-tri-ci-ty,
Good woman that she be.

In the cause of woman's suff(e)rage she shineth as a star,
And as President of Congress she is known from near and far,
For her executive 'bility and for her silver ha'r,
Good woman that she are.

The room echoed with laughter and applause as the rain resumed, slapping against the tin domed roof.

IT WAS EVIDENT THAT Maria Mitchell adored her students, but she did not coddle them. "I cannot expect to make astronomers [of you]," she told them in an introductory lecture, "but I do expect that you will invigorate your minds by the effort at healthy modes of thinking." This was Vassar, after all, and Mitchell demanded rigorous math and meticulous observation. Each clear day at noon, she assigned students to make glass negatives of the sun to trace the movement of sunspots, a painstaking process in a primitive era when photography generally meant preparing one's own chemical emulsions. At night, her students forwent social activities to look for comets, track the moons of Saturn, and count shooting stars when a meteor shower was expected. In the classroom, Mitchell taught spherical trigonometry and the computation of planetary orbits, a skill her students used to write a monthly astronomical column for *Scientific American*. She also taught how to predict eclipses, and in the spring of 1878, the final work for her seniors included a timely challenge: plotting the path of the solar eclipse to occur in July.

Mitchell believed that science was best taught outside the classroom, and with teams of men assembling to go west to observe the solar eclipse, she decided to do something radical—assemble a team of women. She had organized a similar expedition a decade earlier, when she traveled to Iowa for the total solar eclipse of 1869; more than a half dozen current and former students joined her there to assist. But Burlington, Iowa, was not an especially adventuresome destination—indeed, the young ladies from Vassar roomed at the homes of local school friends—and the eclipse predated Dr. Clarke's vexing book that claimed higher education was endangering women's health. At the time, in 1869, Maria Mitchell had not yet emerged as a vocal activist, and she did not seek attention for her unusual entourage of female assistants. On the day of the eclipse, when a reporter asked for an interview, Mitchell declined to comment.

The Vassar eclipse expedition of 1878 would pose a far greater challenge and carry larger stakes; it would be more daring, more distant, more resolutely public. Once again, Maria Mitchell sought volunteers, and she approached some of her most talented students. A natural choice was Harriot Stanton, senior class president and a daughter of Mitchell's friend Elizabeth Cady Stanton, the leading suffragist who had organized the 1848 Seneca Falls Convention in upstate New York that helped launch the women's rights movement. Harriot longed to join the expedition—"What a magnificent opportunity for a young student!" she effused—and one might have expected her mother to support such an overt show of intelligence and independence. But, perhaps because of concerns for her safety, although the stated reason was cost, Harriot's parents told her she could not go. She would always resent the decision. In her memoirs, this daughter of a women's rights icon, who went on to emulate her mother's political activism, called missing the 1878 eclipse "the biggest disappointment of my life."

Mitchell had better luck recruiting recent alumnae. Emma Culbertson had graduated the previous year and clearly viewed Maria Mitchell, in both her career choice and her social activism, as a role model. Culbertson aspired to work in the sciences, not as an astronomer but as a physician (by the 1870s, several female medical colleges had been established, though "lady" doctors remained relatively rare), and at her commencement she had spoken on "Social Prejudices against Woman's Entering the Profession of Medicine." During her senior year, Culbertson helped organize the astronomy class to commission a plaster bust of their esteemed professor, which was later cast in bronze. Another of Mitchell's former students, Cora Harrison, spoke on behalf of the class in offering Mitchell's bust to the college. Harrison had earned her degree a year earlier but stayed on as resident graduate, assisting at the observatory. She was serious about a career in astronomy and possessed her own telescope. It was no surprise that Culbertson and Harrison agreed to join their mentor's expedition.

Two other former students, both of whom had left Vassar in 1873, also signed up for the trip. Cornelia Woods Marsh had studied with Mitchell as part of a two-year, non-degree-granting program and then headed home to Quincy, Illinois. (Mitchell's dome party verse for her began: "Oh C. W. M./ My tears I can't stem/ At the thought of your going away. . . .") Elizabeth Owen Abbot had earned her four-year degree in 1873 and then became a schoolteacher, eventually landing a job in Cincinnati. Mitchell provided a glowing letter of recommendation. "[I] can truly say that I know no better woman, nor better scholar," she had written.

Mitchell invited one more woman to join her all-female eclipse expedition. Phebe Mitchell Kendall, Maria's younger sister by a decade, was a kindred spirit. Though often quiet and shy, she was active in the Woman's Congress and advocated the reform of women's clothing. ("I hope yet to be as simply, comfortably and perfectly dressed as my husband is, and at the same time in a prettier, and less expensive costume," she once confided.) The sisters had been close as girls and remained so in adulthood. The unmarried Maria had accompanied Phebe and her family on a trip to Europe in 1873, and now Phebe, whose son had since grown, would leave her husband behind and accompany Maria to Colorado. Although no scientist, Phebe was assigned an important role. She would serve as an artist, sketching the corona.

Mitchell had now assembled her team for the Vassar eclipse expedition of 1878. "[W]e were a party of six," she wrote, "'All good women and true.'"

CHAPTER 9

SHOW BUSINESS

"Edison exhibits at Irving Hall his new instrument for measuring heat," New York's *Daily Graphic* announced in early June. "As a wonderful inventor, he is himself red-hot." The event, held on a Monday night near Manhattan's Union Square, would offer a public debut of what Edison had first mentioned to the press in April—a hypersensitive thermometer he had created at the request of astronomer Samuel Langley. In advance of the unveiling, Edison's promoter told reporters that they would witness a device of extraordinary powers. "[I]t would require a Fahrenheit thermometer fifteen miles in height to record the same range of degrees of heat," he claimed.

The late-spring evening was pleasantly cool, and New Yorkers were out enjoying theatrical entertainment that ranged from classic tragedy (*Romeo and Juliet*) to modern farce (*Our Boarding House*, featuring a cowardly, eccentric character named Colonel M. T. Elevator). Edison smartly tied his promotional event to entertainment, too. It began with a musical performance that starred his phonograph in concert with a pipe organ and trumpet. As the general public departed, invited journalists were ushered into a separate room to eat a late supper while being fed information about the latest wonder

Reception to Mr. Edison.

After the Concert at Irving Hall, on Monday Evening, June 3, at 10 p.m., a press reception will be given to MR. THOMAS A. EDISON.

You are respectfully invited to be our guest on that occasion.

MR. EDISON *will exhibit his Phonograph, explain his latest invention for measuring the heat of distant stars, etc.*

Supper at 11 p.m.

This Card of Invitation will admit the bearer.

from Edison's workshop. One of Edison's business partners introduced the guest of honor and described the effort required—nearly involving physical force—to drag Edison from his country laboratory. The inventor smiled and obligingly bowed in silence.

Once Edison began exhibiting his heat measurer, however, he had much to say. He described its construction. As in his telephone, the key component was a disk of compressed carbon, set between metal plates and connected to a battery. To this he attached a rod of vulcanized rubber that was set in place so that it pushed against the carbon. "Heat causes the strip of hard rubber to expand and press the plates closer together on the carbon," he explained. "Cold decreases the pressure." In other words, an increase in temperature increased the pressure on the carbon, which boosted the carbon's electrical conductivity and in turn raised the amount of current flowing through the device. The change in current—mirroring the change in temperature—could then be read on a meter, its needle deflected. "By means of this apparatus," Edison said, "it is possible to measure the millionth part of a degree Fahrenheit." (The sensitivity Edison claimed for the device had grown twentyfold in just over a month. When interviewed in Washington, he had said the device could detect temperature changes of one fifty-thousandth of a degree.)

Edison gave only a brief demonstration of the device that evening, but he more fully displayed its capabilities a short time later at Menlo Park. When he held his little finger four inches from the instrument, the needle moved in response. When he held a match six inches away, the needle swung twice as far. He then breathed on the device. The needle deflected three times farther still.

During the press event at Irving Hall, Edison explained how the invention could be used in astronomy. By focusing starlight through a telescope and onto the device, a scientist could measure the tiny amount of heat received from that far-off celestial body. A reporter for *The New York Herald* was impressed. "In this way it is not improbable astronomical researches as to the distance of the stars from the earth may be measured by their degrees of heat," he wrote in the next day's paper. "Indeed, if the experiments made by Mr. Edison last night prove anything they show that he has devised an instrument so sensitive as to open an entirely new field in the realm of scientific research." A reporter for the *Tribune* echoed this enthusiasm but noted that Edison's invention lacked something essential. "No name has yet been given to the instrument," he wrote.

Ensconced back at Menlo Park, Edison scribbled possible names. *Micro-thermo-meter. Micro-thermo-scope. Thermo-micro-meter.* He had recently purchased an etymological glossary of English words derived from Greek. To describe his sensitive heat measurer, he played with various permutations of Hellenic prefixes and suffixes. *Thermo-micro-scope. Carbo-electro-thermopile. Carbon electro thermometer.* In the end, he settled on an especially esoteric name: the *tasimeter* (pronounced ta-*sim*-i-ter), from the Greek for "extension" and "measure," since the device worked by detecting the minuscule growth and shrinkage of the rubber strip. To emphasize the instrument's sensitivity, he sometimes called it a *micro-tasimeter*.

———

IN THAT JUNE OF 1878, Edison was working on a raft of inventions with fanciful names, news of which served as a welcome distraction to a nation relitigating the contested presidential election of 1876 with a congressional investigation into voter fraud. Among Edison's new contraptions was the *aerophone*, which he described as a herculean loudspeaker, powered by steam, that could broadcast the human voice for miles. (Edison suggested installing it in lighthouses to shout warnings at ships.) The *telephonoscope* also projected the voice into the distance, but narrowly—for private conversation—using a megaphone for sending and two ear trumpets for receiving. The *phonomotor* originated as a joke. According to one version of the story, it was during Edison's trip to Washington that a heckler said, "I wonder if you couldn't talk a hole through a board." Never fazed, Edison responded, "Of course, I could," whereupon he sketched a contrivance that, by means of a vibrating membrane connected to ratchets and cogs, harnessed the human voice to turn a drill bit ever so slowly. The same mechanism, Edison said, could wind clocks using the power of conversation. One of his business partners facetiously suggested another application: "It is expected to become a

EDISON'S PHONOMOTOR.

favorite method of suicide, for by its use a man can bore himself to death with his own talk." Meanwhile, during this period of frenzied creativity, Edison was inventing not only gadgets, but a persona, crafting for himself an irresistible public image.

The 1870s was a time of snake oil and flimflam, an era in which the American populace often had difficulty distinguishing scientific fact from pseudoscientific fraud. Séances, at which ghosts purportedly played guitar and rang bells, drew large audiences, and the occultist Madame Blavatsky—derided in her own time as "one of the most accomplished, ingenious, and interesting impostors in history"—convinced followers of the existence of magic and invisible spirits. (Blavatsky was, unsurprisingly, one of the few believers in Edison's mysterious etheric force.) In the medical realm, an 1876 book claimed that exposure to blue light could cure rheumatism, deafness, and baldness, sparking 1877's "blue glass craze," during which merchants peddled cobalt windowpanes to great profit. That same year, an enormous petrified man was unearthed in Colorado and went on display in New York. Nicknamed the Solid Muldoon and touted as a missing evolutionary link, the fossilized creature—which sported simian feet and a four-inch tail—was soon revealed as an elaborate hoax, molded out of meat, bone, eggs, and blood mixed with plaster and baked in a kiln. The underwriter of the enterprise was that shrewd huckster of the age, P. T. Barnum.

Like Barnum, Edison evinced an innate talent for self-promotion, and he too was in show business. Although the phonograph held potential practical value (Edison anticipated its eventual use in recording audio books and teaching languages), it was for the moment a toy, an oddity. The recently incorporated Edison Speaking Phonograph Company saw its best chance for making money through entertainment. It contracted the services of showmen to demonstrate the talking machine before audiences from Philadelphia, Washington, and Atlanta to Indianapolis, Toledo, and Oshkosh. When the phonograph came to town, newspapers listed it under "amusements"; in Chicago, it was up against a show advertised simply as MIDGETS! "[T]he midgets or dwarfs draw $1000⁰⁰ houses while the Phonograph draws $50⁰⁰," Edison's local exhibitor complained. The business needed promotion. Barnum expressed interest in partnering with Edison, but Edison's manager of phonograph exhibitions, James Redpath—who had made his name as lecture agent for such venerable orators as Frederick Douglass and Henry Ward Beecher—strongly and wisely objected. "It degrades the machine into the rank of humbugs for men like Barnum to have any thing to do with it," he cautioned. The way to pitch the phonograph, it seemed, was to stress its genuineness, and that meant selling the authentic genius behind it.

In popular entertainment, a niche existed for authenticity. A common genre on stage was the frontier drama—the predecessor of the Hollywood Western—which featured Indians, outlaws, knife fights, horses, gun smoke, clichéd dialogue, and wooden acting. What elevated these plays from mere trifles was the presence, on occasion, of actual frontier celebrities. Two of the most prominent were William F. "Buffalo Bill" Cody and his friend John B. "Texas Jack" Omohundro, charismatic army scouts and hunting guides who traveled

the West when not on tour in eastern theatres. In June 1878, while the phonograph was on exhibition in Boston, Texas Jack opened across town in a mawkish melodrama called *The Trapper's Daughter; or, Perils of the Frontier.* As one critic remarked, "There is much absurdity in the action as well as in the drama itself, but there is some satisfaction in seeing *bona fide* heroes instead of mock personages on the stage." Yet Texas Jack's

TEXAS JACK.

authentic heroism had recently dimmed. The previous summer, while guiding English hunters on a tour of Yellowstone, he had bragged to the press of a dangerous encounter with hostile Indians. "I received a shot in my stirrup, exchanged shots with them, and returned to my party," he telegraphed *The New York Herald*. But the Englishmen who had hired him soon told the papers a different tale—that their guide was a coward who had gotten them lost in the wilds. A witness to the supposed encounter with Indians claimed that it was Jack himself who had shot the hole through his stirrup. "Instead of being termed 'Texas Jack, the well known scout,'" the man wrote in a letter to the editor, "it should be 'Texas Jack, the blowhard.'" Such was the danger of basing one's celebrity on a claim of authenticity. Critics might try to expose you as a fraud.

Edison seemed determined to avoid that risk. He and his promoters, investors, agents, and partners worked relentlessly to keep the newspapers on their side. Edison flattered and befriended journalists, granting the more obsequious ones liberal access to his lab. In late May, his phonograph exhibitor in Boston arranged a press junket that brought a half dozen reporters to Menlo Park. They were given the run of the place. They could leaf through the library, play with the phonomotor, tour the machine shop, and peer through Edison's borrowed telescope to gaze at the cables rising on the Brooklyn Bridge. To the delight of his visitors, Edison demonstrated his telephone and phonograph and sent each guest home with a souvenir: his autograph elegantly written with the electric pen. The resultant gushing of praise in the Boston papers must have consumed gallons of ink. "ONE OF THE CURIOUS THINGS about the visit was the utter lack of any indication that Mr. Edison felt himself to be a great man," read a typical comment, in the *Boston Herald*. "He chatted about his inventions with as great a degree of freedom from anything like egotism as it is possible to imagine." Edison came across as an approachable genius, humble both in demeanor and in origins, a seductive character in an America that revered the upwardly mobile self-made man.

———

THERE WAS, THOUGH, FAR more Barnum in Edison than Edison let on. He was an unflagging salesman who, whether out of naïve enthusiasm or strategic wiles, effortlessly embellished reality, exaggerating the progress of his work and downplaying obstacles. When he spoke of his latest inventions—the aerophone, telephonoscope, phonomotor—he portrayed them as successful machines even as they remained mere concepts and prototypes. This was true of the tasimeter. "Well, is it perfected yet?" a reporter asked in early June. "It has not got its fine clothes on yet. That's all," Edison replied.

In reality, the tasimeter remained naked. All he had built was the device's central component, which he strung together with other parts on his laboratory bench top to demonstrate the basic principle. As a scientific instrument, it remained far from practical. Yet when astronomers read Edison's earnest comments in the papers, they were eager to try it. Samuel Langley, who had encouraged Edison to create the tasimeter in the first place, was making plans to observe the solar eclipse from Colorado. He hoped to measure how much energy was given off by the sun's corona, which could help reveal the true nature of that mysterious halo. "I expect to go in the beginning of July," Langley wrote Edison in late June. "[I]f the tasimeter arrived this month, I might perhaps be able to put it to practical trial at once in Eclipse work." Other scientists—one from Chicago and a duo from Princeton—also expressed interest in using the tasimeter during the eclipse.

The much-touted invention, however, remained an unproven work in progress. Edison had said that the tasimeter could measure one millionth of a degree Fahrenheit and could detect the heat of a distant star, yet he had provided no evidence to back up either claim. And if the device was anywhere near that sensitive, how would he shield it from extraneous temperature swings? The slightest breeze, even the body heat of the scientist using it, might throw off any meaningful measurements.

Having staked his reputation on a device that did not really exist, Edison now scrambled to bring that device into being. He reconceived the tasimeter's design and had his workmen build this new version. It was small and squat, easily held in one hand. A heavy brass casing surrounded the essential element—the strip of hard rubber that expanded with heat—shielding it from the outside world, except in one direction, where a metal cone projected. The cone looked like the bell of a trumpet, but rather than letting sound out, it funneled heat radiation in. Aim the device through a telescope at a star or the sun's corona, and it would measure the

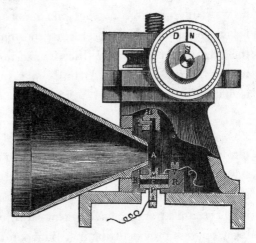

The Tasimeter.

minuscule radiant heat received. At least that was the concept.

Edison, however, had little time to experiment with his new tasimeter; he faced a constellation of other obligations. He was busily improving his phonograph and telephone. He had become mired in a nasty public spat with an inventor in London whose new device for amplifying sound, called a *microphone*, worked on the same principle as Edison's carbon telephone. (Edison accused the man, David Hughes, of "stealing" his invention.) Every day brought a flood of letters from admirers and beggars seeking money, employment, cures for chronic pain and for deafness. And then there were the visitors, many uninvited and some by the trainload, who swarmed Edison's laboratory as if taking in an amusement park. "Mr. Edison says that the bores are coming upon him so thickly," wrote the New York *Sun*, "that one of his earliest inventions in the future must be something by which he shall be protected from their disastrous inroads." He later wished for something else to keep the crowds away. "I have prayed for an earthquake," he said.

It was all getting to be a bit much, even for the seemingly unstoppable Edison. "I am pretty badly used up just now—tired out," he admitted in mid-June, and then fell sick. Newspapers reported that he was suffering from nervous prostration. As one put it, "Edison is laid up for repairs." Concerned friends encouraged him to escape the stress of the laboratory. They invited him to the White Mountains, the Great Lakes, the Atlantic shore. Then came a more alluring invitation.

GEORGE BARKER, EDISON'S FRIEND from the University of Pennsylvania, had never witnessed a total solar eclipse, and he was eager to do so. In April, he had approached Simon Newcomb to see if he might secure a slot on a government expedition to the West, assuming Congress provided funding. But soon after Congress did provide funding, all of the slots were taken. "I have heard nothing from you about the eclipse, and am sorry to find you are not in the list made up at the [Naval] observatory," Newcomb wrote to Barker in early June. Meanwhile, Barker made other plans.

Henry Draper, the New York University professor with whom Barker had observed the transit of Mercury, was assembling a small private eclipse expedition. Although trained as a physician and employed as a chemistry professor, Draper was most passionate about astronomy, especially celestial photography. Even while serving as a

HENRY DRAPER.

young Civil War doctor at the Union garrison in Harpers Ferry, Virginia (now West Virginia), which would soon fall ignominiously into Confederate hands, Draper had found his mind drifting toward the heavens. "This morning I was up in time to see Venus quite distinctly," he wrote his brother in 1862. "She looks splendidly and would photograph easily." Draper's wife, Anna, a striking redhead, shared her husband's passion for astronomy and inherited a fortune from her father, which enabled the couple to equip an impressive private laboratory in Manhattan as well as their observatory in the country. Henry Draper was acquainted with Edison and, while corresponding with him on scientific matters the previous summer, had urged Edison to visit. "I will try and call at your place and see how you peek at the almighty through a keyhole," Edison replied.

Draper's nascent eclipse party included yet another friend of Edison's. Henry Morton was president of the Stevens Institute of Technology in Hoboken, New Jersey, and had organized a government-sponsored eclipse expedition to Iowa in 1869. In 1875, Morton had provided Edison with equipment for his etheric force experiments, and Edison had recently returned the favor by providing Morton with a phonograph. "I am under so many obligations to your kindness," Morton wrote, "that I hardly know how to express my appreciation of your amiability in this relation."

George Barker was then invited to join Draper's eclipse expedition, and the Pennsylvania professor in turn extended an invitation to Edison. It was just what Edison needed: a chance to escape the stress of Menlo Park, far from the parade of curious fans and profit seekers. The expedition also offered Edison another opportunity to put the etheric-force debacle behind him, by spending time in the company of eminent scientists and showing himself to be one of them. In fact, he could bring his tasimeter to conduct his own experiments on the eclipsed sun. "Have seen Professor Barker," Edison wrote to Henry Draper, "and arranged to accompany you as proposed."

The New York Herald broke the news in late June. "The latest marvel from Menlo Park is the 'tasimeter,'" the paper reported. "It is to be used in the scientific experiments to be made by astronomers from all parts of the world at Denver, Col., next month during the total eclipse of the sun, visible at that place. Professor Edison starts for that State on the 8th of July, accompanied by Professors Draper, Barker and other scientists from this section of the country." (Actually, by this point, Draper had shifted his expedition's destination from Colorado—his original plan—to Wyoming, after Simon Newcomb advised that the odds of clear skies were more favorable farther north.)

The article in the *Herald*, one of the nation's most influential newspapers, generated widespread coverage, both excited and fanciful. "Edison is coming West," gushed the *San Francisco Chronicle*. "He will . . . experiment on the moon with his moonograph." "EDISON, the inventor, of phonograph fame, is coming to Colorado ostensibly to witness the eclipse next month, but really with a view to try and connect the other planets with the earth by a telephone wire," mused the *Rocky Mountain News*. More seriously, Philadelphia's *Press* predicted that Edison's decision to go west would do no less than shift the nation's attention. "Now that Mr. Edison has joined the astronomical party to observe the solar eclipse," the paper editorialized, "the event will suddenly be invested with an importance not before dreamed of by the general public. It will at once be concluded that where Edison is, some great discovery will be made."

Indeed, the American people were by now following the inventor's exploits with the same passion they exhibited in devouring serial novels—desperate each month to read the next installment—and there was no chance the public would miss the coming chapter in the adventures of Thomas Edison. An ex-telegrapher friend of Edison's, Edwin Marshall Fox, worked as a journalist for *The New York Herald*. It is not clear who proposed the idea, but a plan was hatched for Fox to join the expedition. He would meet up with the scientists once they settled themselves in the West, then send dispatches back to New York. Edison would travel with his own eclipse correspondent.

———

THE NIGHT BEFORE HE LEFT Menlo Park, Edison once again demonstrated his unrepentant eagerness to please the press. The evening, mercifully mild after a long hot spell, found the inventor reposed in an armchair on the upper floor of his laboratory while an artist, at the request of a magazine, cast his head in plaster. Edison's face disappeared within a heavy, solid block. As the plaster set, he breathed through paper tubes inserted in his nostrils, unable to see or talk. His nephew, who was supporting Edison by the shoulders, felt his uncle grab for his hand. Edison tapped on his palm. It was Morse code. "If I should fall back," Edison wrote, "it will break my damn neck." Removing the plaster proved yet another ordeal. It carried away clumps of hair, leaving the normally unkempt inventor even more disheveled than usual.

It was in this condition that, on Saturday, July 13—five days later than the originally announced departure date—Edison said good-bye to his children and pregnant wife, then headed toward New York to meet his traveling companions. The others—Henry and Anna Draper, George Barker, and Henry Morton—had already shipped nearly a ton of scientific equipment to Wyoming. Edison carried his handheld tasimeter. He had continued to modify its design until two days earlier. With no time to test the device and with the world watching, he did not know if it would work.

Although Edison possessed neither academic credentials nor experience with eclipses, the young inventor attracted the lion's share of press attention. While waiting with his party for the Pacific Express train at the Pennsylvania Railroad depot, he was approached by a reporter who inquired about the journey ahead. Edison could barely contain his excitement. "If the sun's corona has any heat of its own or possesses any heat-reflecting power the tasimeter will measure it accurately," Edison said. "I can hardly wait until I get there. This is the first vacation I have had in a long time, and I mean to enjoy it."

PART THREE

1878

CHAPTER 10

AMONG THE TRIBES OF UNCIVILIZATION

JULY 1878—

Wyoming Territory

IN THE ARID TERRAIN OF SOUTHERN WYOMING, IN A BLEACHED landscape of sage and greasewood, the Continental Divide divides, forming a high basin from which water cannot escape except by evaporating. The region is alkali desert, home to horned toads and rattlesnakes and exceedingly few people. "The eye has no joy, the lips no comfort," a visitor wrote in the 1860s. "[T]he sun burns by day, the cold chills at night; the fine, impalpable, poisonous dust chokes and chafes and chaps you everywhere." Travelers in the nineteenth century generally traversed this wasteland as quickly as possible; it was a hell to be endured, not a destination. But for scientists heading west to view the eclipse of 1878, this was a garden spot, because here, in Wyoming's Great Divide Basin, lay the intersection of two long ribbons across the earth: the transcontinental railroad and the path of totality.

"I arrived here about 12.30 P.M. Laramie time," an assistant astronomer from the U.S. Naval Observatory, A. N. Skinner, wrote to Washington from a remote rail stop on the Union Pacific. "Creston Station is in the midst of an immense elevated plain," he reported.

"The only inhabitants here are Station Master who is a telegraph operator[,] his wife and a few Chinamen." A water tank stood beside the station. Opposite, on a side track, sat the railroad postal car that had ferried telescopes and other scientific equipment from the East. Discarded tin cans and the skull of a cow or buffalo littered the sand. A small graveyard lay nearby. "It is as still as death here," Skinner wrote. "If I had nothing to do I should find it unendurable."

There was much to do, however. Just fifteen days remained before the eclipse, and the junior astronomer was part of an advance team, sent to various sites in Wyoming and Colorado, charged with setting up camp before the arrival of the other, more senior scientists. A local carpenter was already on his way to Creston with a load of lumber to erect a makeshift observatory; its plank sides would shelter the telescopes from the wind, while its canvas roof would allow quick access to the heavens. The U.S. military stood ready to help, too. The army's commander, General William Tecumseh Sherman, renowned in the North but despised in the South for his Civil War rampage through Georgia, ordered logistical help to be provided by Fort Fred Steele, a frontier post that sat along the railroad east of Creston. The fort dispatched a four-mule wagon—it carried a mess tent, a cooking tent, and soldiers to help run the eclipse camp—and would later send fresh supplies of ice and hay.

BY MID-JULY, an impressive assemblage of scientific brainpower never before seen in this remote western expanse was headed toward Creston and other points along the path of totality. Astronomers boarded trains at Boston, Providence, New York, Baltimore, and Washington, and farther inland at Pittsburgh, Cincinnati, St. Louis, and Chicago, for the long trek by rail. "The ride to Harrisburg was excessively disagreeable, being hot, dusty, and smoky," a student from Johns Hopkins complained at the start of his journey to Colorado. Another young astronomer, an associate of Simon Newcomb's, found Pittsburgh "simply dreadful—I've never seen any description of it that I can now pronounce overdrawn—dark, dusty, down in a gully, everything full of smoke & coal-black, & the air of the entire place scandalously vile with coal gas—which almost suffocated me." But soon the landscape opened up—"fields & fields of corn, growing luxuriantly, comprising hundreds of acres, extending nearly as far as the eye can reach."

The westbound scientists joined an eclectic mix of characters, a patchwork of tourists, immigrants, gold seekers, cowboys, politicians, and businessmen. The astronomy student from Johns Hopkins, crossing Kansas, eyed the passengers in his car warily. "There is a pretty rough crowd on board, one fellow opposite with a bag from which he periodically pulls a huge whisky bottle and takes a drink, and another evidently a drover with huge cowhide boots covered with mud, which he rests on the back of the seat before him." Texas Jack—the irrepressible Eastern showman and handsome Western scout—was on another train, heading to Wyoming to guide a hunt for a pair of New Yorkers, one a banana importer; a witness on board noted that Jack "endeavored to get up a flirtation with a very pretty Swedish maiden" but enjoyed little luck, as the girl spoke no English. Simon Newcomb's protégé headed south, toward Dallas, but first had to cross a broad swath of the plains into which the U.S. government had, through coercion and physical force, relocated dozens of Indian tribes. The region would later be overrun by homesteaders and remade into the state of Oklahoma, but in 1878 it was known as Indian Territory, a place both exotic and a bit frightening to the young astronomer from Washington. "[J]ust *think* where I am," he wrote breathlessly to his sweetheart back home, "among the tribes of uncivilization."

OF ALL THE SCIENTISTS heading west, none traveled in more ostentatious comfort than did Thomas Edison. He departed New York in an elegant carriage manufactured and operated by the Pullman Palace Car Company, the firm that set the standard of luxury in nineteenth-century train travel. Pullman cars embodied Victorian opulence: rich upholstery, lavish carpeting, beveled glass, inlays and carvings in walnut, mahogany, and teak. Ventilators provided fresh air free of cinders and dust, while double doors and carefully manufactured windows shut out noise and enabled genteel conversation. Special wheels—made of compressed paper sandwiched between metal plates—smoothed vibrations as the train glided along. At night, hinged chairs folded into beds. Upper berths miraculously descended from the ceiling.

Edison and his companions enjoyed an added level of privacy and splendor. They traveled in an exclusive "hotel car," which featured its own wine closet, kitchen, and chef. At mealtime, waiters transformed the car into an elegant dining room, with fresh linen and china and crystal. A recent passenger on such a posh train effused: "We sip our oyster-soup, discuss turkey and antelope-steaks and

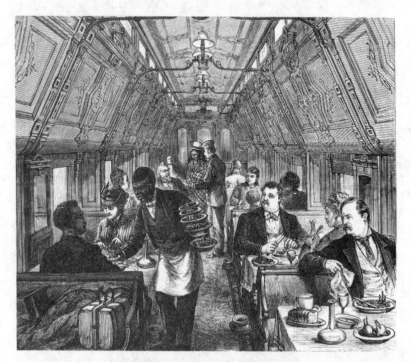

DINING AT TWENTY MILES AN HOUR.

quail, and trifle with ice-cream and *café noir*, with our eyes on the sunset outside."

Such luxurious fare combined with the most modern travel appurtenances continued during a stop in Chicago. Edison and his companions lodged at the Grand Pacific Hotel, a palazzo-style behemoth known to host dignitaries and celebrities, including the ubiquitous Texas Jack on his travels to the frontier. The cavernous, frescoed lobby featured Corinthian columns and a checkered marble floor. A heat wave had settled over the region—sunstroke was felling not only people but also the horses that transported victims to the cemetery—yet Edison, whose overworked nerves had by now attained some degree of repair, donned a straw hat and ventured into the city. Meanwhile, a scrum of reporters staked out the hotel, hoping to snag an audience with the eccentric inventor. When he returned that evening, they descended en masse.

"Have you ever been in Chicago before?" one newsman asked.

"Yes," Edison replied, "thirteen years ago. I had a linen duster, $2.50, and a railroad pass. I was not interviewed then."

"How many patents have you now?" queried another.

"It was something like 140 some time ago. I guess there are 150 or so now."

The journalists wanted to hear of Edison's phonograph, his plans for the eclipse, and his latest scientific invention.

"The tasimeter is a heat-measurer of very delicate power," Edison explained. "Now, it is very evident that the sun's rays, those coming from the center, are hot—quite hot, as you have found them here lately."

At this the reporter, as the scribe himself described it, "perspired assent."

"It isn't so evident that there is great heat in the rays near the sun's edges," Edison continued. "Now, when the moon crosses the sun, all but the edges of the [sun] will be covered for two minutes fifty-seven seconds. These edges, or the corona, are what we want, and the tasimeter, with its wonderful delicacy, will measure the heat."

"You are hopeful that the instrument will do what you claim for it in the eclipse?" a reporter pressed.

"Yes; I don't see why it shouldn't," Edison replied, turning on his folksy charm. "I'm no astronomer myself—don't know anything more about it than a pig does about learning Latin. What I want to see is, whether my instrument will do the work, and I guess it will."

AS EDISON JOURNEYED WESTWARD, the torrent of flattery and attention continued. At Omaha, where he and his entourage transferred to the Union Pacific, he was confronted by another journalist with notebook and pencil. "Well, then, begin your fusillade of conundrums, and I'll give you all the taffy you want," Edison offered. "I am used to it. My place at Menlo Park is headquarters for the New York reporters." He seemed content to talk all day, until Henry Draper broke in: "Come on, Tommy, let us go down to the train." (Some of Edison's friends called him by a variant of his first name. Others called him "Al," for his middle name.) Before embarking, the inventor-turned-celebrity received a perk from the railroad—a special pass that permitted him and his party "to ride on the Locomotive or where else they may desire."

Edison's desire, often, was to perch on the engine's prow. He lounged on the cowcatcher, on a cushion provided by the engineer, propelled forward by iron and coal and steam as he took in the scenery "without dust or anything else to obstruct the view," as he put it. Nebraska displayed a vast, subtly shifting panorama: Cloud shadows on the broad plains. The lazy Platte River. Clumps of cottonwoods. Prairie dog towns. Cattle, where bison recently had roamed. The hypnotic passage of the telegraph posts that stretched to infinity. Edison found his bliss marred just once, when the locomotive struck an animal—a badger, as best he could tell. He grasped the angle brace and hung on tight as the train batted the creature into the air.

Finally entering the territory of Wyoming, the train panted up a long grade, past enormous outcroppings of pink granite that emerged

CROSSING DALE CREEK BRIDGE—130 FEET HIGH.

from the prairie in jumbled, bulbous piles. Soaring mountains rose on the horizon, their tops still adorned with snow in midsummer. "Saw Pikes Peak. 160 miles away," recorded Henry Morton, who joined Edison on the cowcatcher. Soon the track passed over the Union Pacific's highest span, a spindly trestle that traversed a gaping ravine in the bottom of which ran a minor stream. The crossing was, in the words of a railroad construction engineer, "a big bridge for a small brook that one would easily step over," but the dizzying view down to Dale Creek, 130 feet below, invariably elicited gasps from those onboard. The train then veered north over flat terrain to skirt the northern end of the Medicine Bows, an imposing rampart of glacier-carved peaks to the west. As night fell, high plains turned to high desert. Buffalo grass gave way to sage.

Edison presumably reentered the passenger car by the time it reached Percy station, near the base of Elk Mountain, its broad, conical form looming like a volcano. Here, seven weeks earlier, the Gilded Age had encountered the Wild West in frightening style. On that day, when the same westbound train—the No. 3—left Percy at 10:00 P.M., four men in white masks clambered aboard the middle

Pullman sleeping car and cried "Hands up!" Two of the criminals stood guard while the others walked the aisle and forced the male passengers, at gunpoint, to surrender watches and cash. The ladies were reportedly treated with more courtesy, taken at their word when they claimed to have no money. Although no one was seriously hurt, the thieves escaped into the hills with their loot, as well as a chunk of the railway's reputation. ("Four Masked Men Clean Out a Palace Car on the Pacific Railroad," *The Philadelphia Inquirer* announced in a front-page headline.) The Union Pacific sought quick justice. "The company will pay $1,000 for the capture of the robbers—dead or alive," the railroad's division superintendent, Ed Dickinson, immediately declared. A posse soon captured the men and placed them in jail, where they awaited trial.

The No. 3 encountered no such trouble on this day, however. It arrived safely at Edison's final destination around midnight. Superintendent Dickinson met the scientific party at the depot.

RAWLINS, WYOMING—POPULATION SIX HUNDRED—sat among treeless hills just east of the Great Divide Basin. It was a railroad town, established a decade earlier for construction of the Union Pacific, and the company still dominated the place—controlled its economy, dictated its daily rhythms, filled its streets with the rumble and hiss and squeal of the trains. Edison, Barker, Morton, and the Drapers descended into darkness and walked along the platform to the Railroad Hotel. A few nights later, Edwin Marshall Fox of *The*

RAWLINS—A COMMUNITY OF RAILROAD EMPLOYÉS.

New York Herald joined them, having telegraphed ahead—"I will be at Rawlins on Sunday night & suppose have you got a bunk for me[?]" Fox shared a room with Edison.

The hotel, a simple two-story frame building on the south side of the tracks, boasted "good fare and commodious rooms"—thirteen in all—plus a large dining area that hosted jamborees, charity balls, and concerts. Other forms of entertainment could be found outside the hotel, in the town's saloons and brothels, where men more typically squandered their evenings on liquor, women, gambling, violent altercations, or a combination of the above—as occurred at a cathouse called the Myrtle Bower, where a prostitute, for reasons unstated but easily imagined, stabbed a young man in the side and shoulder.

This boisterousness occasionally infiltrated the Railroad Hotel. One night, Edison and Fox awoke to a thunderous knock. They opened their door to find a handsome, buckskin-clad fellow, drunk and armed. The intruder said he wanted to meet the illustrious inventor, whom he had read about in the newspapers. He then pulled out his Colt revolver. Boasting that he was "the boss pistol-shot of the West," he pointed out the window at a weathervane on the freight depot, then aimed and fired, hitting his mark. "The shot awakened all the people, and they rushed in to see who was killed," Edison later recalled. Edison was understandably unnerved, but the gunslinger proved to be no outlaw. It was none other than Texas Jack. History does not record his side of the story, but one gets the sense of waning stardom, a performer who felt compelled to prove his mettle and manhood to America's newest Gilded Age icon. (Texas Jack was reportedly deep in debt at the time, and his marriage was suffering.) A flummoxed Edison tried to defuse the situation. He said he was tired and would see Texas Jack in the morning. The bloodshot visitor finally left. "Both Fox and I were so nervous we didn't sleep any that night," Edison reported.

In daylight, Rawlins revealed itself to be drab and rough-hewn, a town the color of rust. The homes, the fences, the depot, the train cars—most everything was coated with "Rawlins Red," a paint pigmented with iron oxide mined from local sandstone. A recent visitor had noted the lack of greenery. "Rawlins presents to the curious eye the usual features of severely utilitarian frame houses, the absence of flowers, turf or shrub, and even of a chance or premeditated inclosure where such might be nourished."

Edison roamed the dirt streets. The small downtown, which straddled the railway, included merchants of dry goods, liquor, clothing, and jewelry, as well as a lumberyard that moonlighted in coffins. Curious about life in the frontier West, Edison and Fox ambled up to the corner of Fifth and Cedar to visit the county jail, a one-story stone structure with three small cells. Incarcerated at the time were a horse thief and three of the four men accused of robbing the Pullman car near Percy station. (The fourth had turned state's witness and was being held separately, in Laramie, for his own safety.) Edison entered, and through thick bars he observed one of the accused train robbers. "He looked like a 'bad man,'" Edison said years later. "The rim of his ear all around came to a sharp edge and was serrated. His eyes were nearly white, and appeared as if made of glass and set in wrong, like the life-size figures of Indians in the Smithsonian Institution. His face was also extremely irregular." Edison tried to initiate a conversation. The prisoner declined to speak.

Most everyone else in Rawlins, however, showed great hospitality. Ed Dickinson, the railroad superintendent, acted as tour guide, taking Edison, Fox, and Morton on an afternoon excursion to Brown's Canyon, twelve miles outside of town, and helping the New Jersey inventor to purchase fishing and hunting gear, including a thirty-five-dollar Winchester rifle, from a merchant in Laramie. Another railroad man, Robert Galbraith, who oversaw the Union Pacific repair shops in Rawlins, offered his house and yard to the eclipse party as a base of operations. William Daley, a carpenter who owned the Rawlins lumberyard and had overseen construction of the Railroad Hotel and the county jail, consented to build for the scientists a pine shed with a removable roof to house their telescopes. (It was Daley who also constructed the crude observatory for the U.S. government astronomers up the tracks at Creston station.) Meanwhile, Lillian Heath, an inquisitive twelve-year-old who would later become Wyoming's first female doctor, befriended the scientists. They allowed her to look through one of their telescopes.

Edison also met Nathan Meeker, a novelist, journalist, and founder of a Utopian agricultural community northeast of Denver. To pay off debts, Meeker had taken a new job as Indian agent on the Ute Reservation, which encompassed much of the western third of Colorado, south of Rawlins. It was Meeker's task, on behalf of the federal government, to convert the Indians from equestrian hunters to sedentary farmers. Meeker was naïvely confident that he would succeed. He invited Edison to go with him to the reservation, "to see the red man on his native heath." Edison declined.

Edison accepted, however, another invitation to venture into the mountains south of town. After the eclipse, he went camping with a party of railroad men accompanied by Major Thomas Tipton Thornburgh, commanding officer of Fort Fred Steele, a skilled marksman whom the Sioux had taken to calling "The-chief-who-shoots-the-stars." Edison, less able with a rifle, merely shot sage hen and deer, and he fished. "You could throw a crumb of bread into the water, and the river was in a froth right away with trout after that crumb," he remarked. "If you held a hook without any bait a foot above the water the trout would leap at it." The men ended up with more fish than they could handle. They ate the best parts and threw the rest away.

Edison relished the wildness and adventure of the American West, but beneath the romantic patina lay a foundation of ignorance, lawlessness, and brutality. Nathan Meeker—the idealistic Indian agent—would be dead a year later, slain in a massacre by the people he had pledged to protect yet whom he came to resent and antagonize when they refused to change their culture in the face of his moralizing and threats. (Meeker would be found with his skull crushed and a barrel stave driven through his mouth, to keep his

DEATH OF MAJOR THORNBURGH

tongue from telling lies in the afterlife.) The ghastly slaughter and a related battle between the Utes and the U.S. Army claimed the lives of dozens: civilians, Ute warriors, and white soldiers, including Edison's own fishing companion, Major Thornburgh. Meeker's wife and twenty-two-year-old daughter survived the ordeal but endured three weeks' captivity during which, they claimed, they were "outraged" by their male captors. The marauding Indians also suffered an unspeakable fate. Although few Utes were involved in the massacre, the entire tribe lost its ancestral homeland when Colorado's political leaders, portraying the Indians as bloodthirsty savages, convinced Washington to force the Utes off their reservation and give the state's white population what it had long coveted—millions of additional acres to mine, to ranch, to settle.

During this era, the good people of Rawlins exhibited their own thirst for blood. One year after the Meeker Massacre, the county jail held a bandit (not among those visited by Edison) whose gang of outlaws, after failing to derail and rob a Union Pacific train, murdered a deputy sheriff and his partner who were in pursuit. The prisoner—he went by the alias "Big Nose George" Parrott—tried to escape, and in the process wounded the jailer. Masked towns-men took matters into their own hands, vigilante style. That night they stormed the jail, hauled Big Nose George down to the railroad tracks, and tossed a hemp rope over the cross arm of a telegraph pole in front of Fred Wolf's saloon. A crowd assembled, comprising "some of the best people of the town[,] tax-payers and law-abiding citizens," a witness reported. "[A]ll seemed to be fully satisfied with the lynching of the prisoner."

The lynching did not go smoothly. Big Nose George stood on a barrel, the rope around his neck. Someone yelled, "Kick the barrel," which was done, but the rope was too long, or slipped, and he fell, painfully alive, to the ground. "Hang him over and make a good job of it this time," was the call from the crowd. Now a ladder was placed against the pole. "Give me time and I will climb the ladder myself, and when I get high enough I will jump off," the doomed man pleaded. Instead, as he mounted the ladder, someone pulled it from beneath. His hands came untied, and he grabbed at the pole, clung to it, climbing several feet with the noose around his neck. "For God sake, someone shoot me," he cried. He slipped down the pole, still hanging on. "Do not let me choke to death!" Eventually, though, he did.

William Daley—the coffin-selling carpenter who had built the Draper party's rough observatory and also served as an undertaker—cut the body down and stored it overnight. The next day, railroad mechanic Robert Galbraith—who had allowed Edison and his compatriots to use his house and yard—served on the jury for the coroner's inquest, which claimed it could not identify the vigilantes, calling them simply "a party of masked men to us unknown." Lillian Heath—the scientifically minded girl who had befriended Edison's band of visiting astronomers—now fifteen, precociously assisted with the autopsy. The doctors in charge eventually pickled the out-law's corpse and stored it in a whiskey barrel for experimental dissec-tion, but first they offered Heath a keepsake. She took the skullcap, still bloody and with hair attached, which in later years she used as a crude doorstop and flowerpot. One of the doctors, meanwhile, harvested a souvenir for himself. He flayed Big Nose George's chest, and had the skin tanned and fashioned into a pair of Oxfords. The doctor later became governor of Wyoming. It is widely reported that he wore those shoes to his inauguration.

CHAPTER II

QUEEN CITY

JULY 1878—

Denver, Colorado

Two hundred fifty miles southeast of Rawlins, down the spine of the Rockies, stood a frontier community of a different sort. Denver was no rude, utilitarian borough, but instead an oasis of civilization in a barbarous West. Though just twenty years old, this booming city near the base of the mountains already boasted an opera house, three hospitals, eight banks and as many newspapers,

LARIMER STREET, DENVER.

twenty houses of worship, forty-seven hotels and boarding houses, and more than twenty thousand souls. Whereas Rawlins had been created to serve the railroad, Denver had created railroads to serve itself. The city's early leaders, intent on turning their undistinguished settlement into the region's economic hub while amassing personal fortunes for themselves, arranged for rail connections toward seemingly every point on the compass. In the downtown business district, stout brick buildings lined a neat grid of streets. In residential areas, such as the confidently named Capitol Hill, where a large undeveloped parcel was the only evidence that Coloradans hoped someday to build a statehouse there, dignified homes sported mansard roofs and cupolas. Horse-drawn trolleys plied the broad avenues, past lush yards with shade trees and irrigated lawns, while municipal water and gas coursed underground. Although Denver lacked refinement in obvious ways—its unpaved streets turned to mud in the rain, and the stench of sewage often floated up from the South Platte—the city aspired to elegance, even enlightenment. "Let the echo go out to the country that the 'Queen City of the Plains' is in fact a city of progress," crowed the *Denver Daily Times*.

Denver attracted rugged doers: merchants and mining engineers, ranchers and blacksmiths, architects and carpenters. The population was largely male and white, and these white men clearly dictated the city's affairs, but others found opportunity here as well. African Americans became barbers, firemen, and hoteliers; the owner of Denver's successful Inter-Ocean Hotel was a former slave. Chinese laborers, laid off after building the transcontinental railroad, opened busy laundry establishments (although they also endured simmering xenophobia that would boil over, in 1880, in a Chinatown-destroying riot, and, two years later, federal passage of the Chinese Exclusion Act). Unmarried women sometimes toiled as servants or prostitutes, but others thrived as professionals and entrepreneurs, opening millinery shops and medical practices. People frustrated or tarnished in the East moved to Denver to reinvent themselves.

Many also came for the climate. The region's dry, fresh air was said to be a tonic for respiratory ailments; tuberculars and asthmatics made up a significant proportion of the population. "Why, Coloradoans are the most disappointed people I ever saw," quipped P. T. Barnum, who owned property in the state. "Two-thirds of them come here to die and they *can't do it*." The people of Denver and environs did not lack for braggadocio, and that pride extended to the astronomical event about to take place. "Sir, Colorado can beat the world in eclipses as in everything else," a local proclaimed to a visitor from England.

As the most populous American city in the path of totality, Denver was to become the main destination for many eclipse chasers. "Tourists are coming into Denver so thick that they can hardly find hotel accommodations," one newspaper reported. "It looks as though by the time the eclipse gets around there will be no longer room in-doors, and that tents will be the only resort of the more tardy visitors." Overflow guests sought rooms in private residences, while others slept on cots in hotel parlors and dining rooms, and one reportedly bunked on a billiard table. The visiting hordes included prominent citizens—newspapermen, financiers, judges, U.S. senators—and with them came a less desirable lot: at least thirty assorted pickpockets, till tappers, and other petty thieves from New York, according to Denver police. But the most notable visitors were the astronomers, who grew as "thick as blackberries" in Colorado. "A cannon shot fired in any direction from a central point in that state would, it is thought, cause a sad falling off in the number of college professors who expect to report for duty at the beginning of the fall term," commented the Washington, D.C., *Evening Star* a week before the eclipse.

The largest group of scientists, and the first to appear, arrived from stately Princeton. Three professors and seven former students, plus one professor's wife and another's son, stepped off the Kansas Pacific Railway in the first week of July. After a few nights at the Grand Central Hotel, which advertised a commanding view of the mountains for just three dollars per night, they pitched tents in a cottonwood grove a few miles southeast of downtown. The camping ground sat sandwiched between Cherry Creek and a smaller stream, the water serving as a partial moat to prevent cattle from wandering in at night and disturbing the instruments. These scientists erected a cookhouse and a privy, a darkroom for photography, and sturdy piers on which to mount their telescopes and spectroscopes. On a rope above the camp's entrance they suspended the Stars and Stripes, flanked by two orange-and-black college flags. Although bothersome flies and mosquitoes presented a nuisance, "Camp Nassau" was an otherwise comfortable place to spend a few weeks. Of course, it helped to come from a privileged and moneyed eastern institution, and to have the conveniences of a city next door. The resourceful Ivy Leaguers hired a black cook to make their meals, and at least one professor noted in his diary that he relied on a Chinese launderer to do his washing.

The lead astronomer at the Princeton camp was a man both well-admired and well-liked in the scientific community. Known affectionately to his students as "Twinkle," Charles Young exhibited a kind and winning personality, traits he had used to extract a favor from Menlo Park—a tasimeter, which he brought to Denver from New Jersey. The amiable astronomer had previously, at the eclipse of 1869 in Iowa, won the hearts of Maria Mitchell's young assistants. "We were all charmed with Prof. Young, whose modesty, tho' he was beginning to be famous, was in striking contrast to the 'sirs' of some of the other masculine scientists," recalled an alumna long after.

CHARLES A. YOUNG.

Under Young's genial leadership, the Denver Princeton camp became a social center, a place where many prominent scientists dropped by and one—British astronomer Arthur Cowper Ranyard—stayed. The Englishman and the team of Americans worked long hours, yet took time to enjoy the natural setting: building a bonfire, admiring clouds at sunset. They also engaged with the locals. One evening, it was a high school astronomy class that visited the camp. Another time, it was a band of Utes, whose leader, given his large size and angry countenance, was feared by many settlers. Chief Colorow was, with reason, bitter about white intrusions on

COLOROW

traditional Indian lands, and in 1879 he would help command the battle that killed Major Thornburgh, but at the time that his party visited Cherry Creek in 1878, it was merely to rest and gather wild plums. "The [Princeton] students were especially interested and talked to Colorow about the coming eclipse," according to a later account by one who was there. "[T]he chief was skeptical and showed it by much grunting. Even if the thing were going to happen, how did they know it?" Before long, the Utes mounted their horses and headed onto the plains in search of buffalo.

MANY DENVERITES INQUIRED ABOUT the eclipse—precisely when it would occur, how to view it, why astronomers were so keen to study it—and the frontier press sought to provide answers. "The eclipse will be total here for the space of 2 minutes and 46 seconds," wrote the *Denver Daily Times*, adding that totality was predicted to begin on July 29 at about 3:30 P.M. The article mentioned that the solar corona "will be the chief object of study of most of the scientific parties who will observe the eclipse from this locality." Other articles ventured to describe what the sky surrounding the corona would look like, and where one might find the hypothetical planet Vulcan. "At the time of the eclipse the star Procyon should be visible at quite a distance from the sun, almost directly below him, Venus considerably to the right of Procyon, [Castor] and Pollux above Venus, and Mars and Mercury above the sun and further to the left, with the star Regulus (the handle of the sickle in Leo) between them," explained the *Rocky Mountain News*. "[I]f any new planets are discovered, they will be nearer the sun than Mars."

Some Coloradans were eager to help the professional astronomers by making detailed studies themselves. "[I] will endeavor to show my *patriotism* and *interest in the advance of science* by taking observations from some point," a Reformed Episcopal minister in Boulder wrote to Simon Newcomb in early July. "A few of our citizens have interested themselves to attend to the matter here—and doubtless others will in other parts of our state." In Denver, several citizen scientists volunteered to assist astronomers on the day of the eclipse, and a visiting team from the Chicago Astronomical Society devised a way to harness the energy of the locals through a coordinated study of the solar corona.

Though often described as a halo, the corona is not a simple, uniform ring around the sun, but a textured, frilly, dynamic radiance that was extremely difficult to photograph using nineteenth-century technology. The human eye can perceive far more detail and a much greater range of brightness than cameras of that era could record, and therefore, at the 1878 eclipse, accurate drawings of the corona were deemed especially valuable. Totality would last, however, less than three minutes, hardly enough time for even an accomplished artist to make a thorough sketch. Elias Colbert, director of

Chicago's Dearborn Observatory, aimed to address this problem through an early form of crowdsourcing. He put out the call for Denverites to participate in a joint corona-drawing exercise, and he invited volunteers to attend a class at the high school. Twenty people showed up—men and women alike, including a teacher, a jeweler, a clerk, and a surveyor. Colbert stood before a blackboard and drew a rough sketch of the corona. He indicated key features to look for: the corona's overall size and shape, whether its rays were straight or curved, whether they extended all the way down to the sun's surface. He then divided the class into teams and the corona into quadrants. On the day of the eclipse, each team would be tasked with drawing, in as much detail as possible (and with the aid of opera glasses), just its assigned piece of the whole. Members of most teams would draw one quarter of the corona; those on another team would ignore the corona and concentrate on the flame-like protuberances at the edge of the sun. After totality, Colbert would combine these partial images into a full, composite sketch that he hoped would become the definitive image of the total eclipse as seen from Denver in 1878.

In their workaday lives, the people of Denver, like those of the rest of America, spent most of their time with their heads down, focused on earthly affairs of commerce and production, but even this go-ahead city saw reason to pause for what was about to happen overhead. "Many persons went down to their graves at the ripe old age of three score and ten, without witnessing so sublime a spectacle in nature as a total obscuration of the sun's disk by the moon," wrote a local correspondent in the *Rocky Mountain News*. "The people of Colorado may thank fortune they are in a position to see this great event in nature to the best possible advantage." The week before the eclipse, the *Denver Daily Times* offered a suggestion: "Wouldn't it be an excellent idea for our business men and bankers to close their places of business next Monday from 1:30 to 4 o'clock p.m., giving employers and employes [*sic*] alike a chance to witness that which has brought visitors from all parts of the world to Colorado."

Of all the visitors coming to the frontier, one in particular continued to transfix the locals. "His name is Edison, the great inventor, who declares he knows no more about astronomy than a pig does about learning Latin, and that he comes West to try one of his little inventions which he calls the tasimeter," the *Denver Daily Tribune* reminded its readers. As late as mid-July, even after Edison had arrived in the region, the people of Colorado had not yet learned that the famous inventor had shifted his destination from Denver to Rawlins. Rumors circulated that he would join the Princeton party on Cherry Creek. The Denver Press Club planned a special reception to welcome its celebrity guest. Then, on July 21, under the headline EDISON'S GO-BY, the *Tribune* revealed the disappointing news. "After all the expectation which has been created on the subject, it has now been affirmed and the statement confirmed that Professor Edison will not visit Colorado for the purpose of observing the coming solar eclipse. THE TRIBUNE regrets this as much as anyone can, because we were anxious that Denver should enjoy the distinction that such a visit would have given it."

Yet Denver could still claim a celebrity. "Mr. Edison is doubtless the most famous inventor of this or any other age," the *Rocky Mountain News* commented, "but we doubt whether he deserves more credit for his marvellous attainments in invention than does Maria Mitchell for demonstrating the capacity of women for the highest and best mental activity and scientific research."

THE VASSAR ASTRONOMER reached Denver on Wednesday, July 24—a week after Edison's arrival in Rawlins but still five days before the eclipse. The journey had been long and trying. Mitchell started in Boston with sister Phebe, who lived in Cambridge, and the pair then headed eight hundred miles southwest to meet Elizabeth Abbot, class of 1873, in Cincinnati. Now a threesome, they continued westward, "hour after hour and day after day . . . over level, unbroken land," Mitchell recalled of the scenery outside. She also observed the scene within the train. "One peculiarity in travelling from East to West is, that you lose the old men," she had written on a previous journey to the Mississippi Valley. "In the cars in New England you see white-headed men, and I kept one in the train up to New York, and one of grayish-tinted hair as far as Erie; but after Cleveland, no man was over forty years old."

As for women, it was not uncommon to see them on western railroads, but they generally traveled with husbands. Riding trains, especially alone, proved more difficult for female journeyers—lugging heavy trunks, trying to look presentable after a night in a sleeper car. A friend of Mitchell's, en route to Colorado in 1880, noted the flagrant inadequacy of facilities for ladies. "Thirty-three women and children and two men used our dressing-room to-day, the latter entirely without right," she grumbled. By some accounts, rail travel was said to be—like university studies—unhealthy for women, the jostling and fatigue taking its toll on the reproductive organs. According to a doctor who studied the matter, a woman who made a habit of riding trains did so "at the expense of her future usefulness."

Mitchell and her companions took a break from the rails and stopped for the night in Kansas City, a major transfer point for both people and animals. As the women prepared to continue west, bellowing cattle were heading east, funneled through the stockyards on their way to distant markets. Here the eclipse party again increased by one—former student Cornelia Marsh joined the expedition—and then it was on again toward Colorado, across the plains where settlers were literally carving homes out of the earth. Even in this land of sod houses and dugouts, well outside Vassar's cultural sphere, Maria Mitchell's name was known. Indeed, to some it was an inspiration. "I am thirteen years old next January and live far out on the frontier, between forts Larned and Dodge," a Kansan—a boy—wrote to her a few years earlier. "Please tell me what books I must get in order to make myself a thorough

Astronomer. I should like to know whether there is an inter-mer-curial planet named Vulcan."

It was twenty-eight hours from Kansas City, on the Missouri River, to the Colorado city on the Arkansas where the Vassar corps was to transfer for the final leg of its trip. Here, in Pueblo, the women encountered a snag. Rail passengers in the nineteenth century faced not only the risk of banditry and accidents but the inconvenience posed by inter-company squabbles, and the railroad Mitchell's party had just left—the Atchison, Topeka & Santa Fe—was engaged in a nasty fight with the line they were about to board—the Denver & Rio Grande. The two roads had been racing to extend their service through the strategically important Royal Gorge of the Arkansas River to the silver mining district of Leadville, Colorado. This battle was being fought in the courts and on the land, where the compet-ing companies sabotaged each other by cutting telegraph lines and throwing the tools of rival work crews in the river, and it would soon escalate to the amassing of weapons and paid gunslingers in prepa-ration for armed conflict.

Mitchell lamented, "We learned that there was a war between the two railroads which unite at Pueblo," and she was inextricably caught in the middle. The Denver & Rio Grande was refusing to carry freight and passengers under the terms of its contract with its rival. The Vassar women were told that there was "trouble" with their through-tickets. Mitchell, apparently, would have none of it. "[W]ar, no matter where or when it occurs, means ignorance and stupidity," she complained. Somehow she got her party on the con-necting train for the final, five-hour journey north, along the foot of the Rockies. Debarking at last in the Queen City of Denver, Mitch-ell and her three companions met the two remaining members of their group—Cora Harrison '76 and Emma Culbertson '77—who had arrived earlier on their own.

The women had landed not just in the West but in a cauldron of gender politics. As a Denver correspondent for the New York *Sun* remarked of the all-female expedition: "This party adds peculiar interest to the work of observing this eclipse, for it is here that woman is making a heroic struggle for equal rights."

THE PREVIOUS AUTUMN, just a year after achieving statehood, Colorado had considered a radical proposal, something not yet done by any state (although it had been done by the territories of Wyoming and Utah)—granting women the right to vote. The legislature had put the question of female suffrage on the ballot, sparking a fierce battle that attracted national attention. Among the notables who barnstormed Colorado by stagecoach, train, and burro was the indefatigable Susan B. Anthony—a close ally of Elizabeth Cady Stanton and long a leader of women's rights campaigns—who visited mining camps and saloons in an effort to persuade the state's men to support the measure. She argued that denying women the vote was un-American, amounting to taxation without representation. Conservative clergy preached the opposing side, claiming that extending this right to women would ruin the family and weaken society. "How absurd and revolting to think of a woman leaving her household duties and abandoning her family to go to the polls, to attend political meetings, to suppose that she can be elected for sheriff or constable!" proclaimed the Right Reverend Joseph Projectus Machebeuf, Denver's first Roman Catholic bishop, a man greatly respected for bringing churches, schools, hospitals, and religion to the frontier. "But what sort of woman are the leaders of these pretended Woman's Rights?" he asked rhetorically from the pulpit. "Some old maids disappointed in love; wives of what are called hen-pecked husbands; women separated from their husbands, or divorced by man of a sacred obligation imposed by Almighty God." The bishop and his ilk predictably won the day—the measure failed by a two-to-one margin—but proponents did not give up. Instead, they braced for a protracted fight. "We must 'keep pegging away,'" declared the president of the Colorado Woman Suffrage Association, Dr. Alida C. Avery.

Avery, the daughter of a New York abolitionist, had long emulated her father's social activism, and she was an old colleague of Maria Mitchell. At the founding of Vassar Female College, when Mitchell was hired to teach astronomy, Avery came on board as the school's resident doctor and professor of physiology and hygiene. Avery could be stern—students feeling poorly were said to "fear the physician more than the impending illness"—but she and Mitchell became allies, fighting side by side for women's rights. Together, they denounced Dr. Edward H. Clarke's book that claimed higher education was turning American girls sickly and sexless. ("I grind my teeth in despair over it," Avery wrote.) Together, they served in the leadership of the Woman's Congress. And, together, they pressed Vassar's board of trustees to provide equal pay to male and female faculty, even threatening to quit over the matter. Avery did quit, and, in 1874, relocated to Denver, a far remove from the college's internecine politics. She opened a private medical practice and soon earned a sizeable income—ten thousand dollars a year, by one account—as well as a solid reputation. In 1879, Avery would be called upon to treat Josephine Meeker, daughter of the slain Indian agent Nathan Meeker, after the young woman's release by Ute captors.

Despite moving to Colorado, Avery remained in close touch with her friend at Vassar. She and Mitchell corresponded by mail and saw each other occasionally at women's rights meetings. On one of Avery's visits to the East, Mitchell mentioned the upcoming eclipse and her desire to observe it. Trying not to be presumptuous, Mitchell asked her friend, "Have you a bit of land behind your house in Denver where I could put up a *small* telescope?" Avery's reply: "Six hundred miles." (Denver's backyard stretched six hundred miles across the open plains to the Missouri River.)

Avery's personal patch of land was considerably smaller. She lived in a gracious two-story home surrounded by a rose garden, a grape arbor, apple trees, and elms at the corner of Twentieth and Champa Streets, where horse-drawn streetcars rattled past. As Denver's prosperous lady physician, she often hosted female

Dʀ AVERY'S RESIDENCE 20ᴛʜ St. DENVER, COL.

intellectuals. A frequent visitor was Helen Hunt Jackson, a much-celebrated Colorado essayist and poet—and a childhood friend of Emily Dickinson—who would soon make her cause the rights of Native Americans. (Jackson's novel *Ramona* would open America's eyes to the maltreatment of Indians much as Harriet Beecher Stowe's *Uncle Tom's Cabin* had awakened the nation to the cruelties of slavery.) Susan B. Anthony, an activist of a different sort, spent three weeks in Avery's guest room after her pro-suffrage speaking tour of Colorado in 1877. Anthony admired the home's immaculate Victorian decoration, with china, silver, cut glass, and all manner of vases filled with fresh flowers. "[I]t is just lovely," she wrote to Elizabeth Cady Stanton during her relaxing stay. "I have it charmingly *free* & easy." The setting inspired Anthony to write a speech called "Homes of Single Women," which praised the elegant residences of America's new class of unmarried female professionals.

Avery now hosted the Vassar eclipse party, and she undoubtedly hoped that her new guests would also serve a political purpose, to show the people of Colorado that women's-rights women could be intelligent, strong, and—bowing to the conceits of the day—feminine, contrary to the portrayals by Bishop Machebeuf and Dr. Clarke. Her new houseguests, however, would not find their stay as comfortable as Susan B. Anthony's had been.

Maria Mitchell, before departing Boston, had packed two telescopes. To prepare them for the journey, she had removed the lenses from the tubes and carefully placed everything in her luggage. But once she reached Denver, she discovered that just the trunk containing one telescope tube had arrived with her; the luggage containing the other tube and the lenses had gone astray. The Vassar women spent desperate hours at the Denver & Rio Grande Railway depot, nine blocks from Avery's house, berating agents and sending telegrams, as they attempted to locate their baggage. With just a few days until the eclipse, they frantically explored whether anyone in Denver had a telescope for sale or for lease.

The moon—an ever-shrinking crescent in the dawn sky—orbited the earth at more than two thousand miles per hour as it inexorably headed toward its rendezvous with the sun. Mitchell's bags, however, sat immobile 120 miles to the south, in Pueblo, where they had become prisoners of the railroad war, putting the Vassar expedition's success—both scientific and political—in jeopardy.

CHAPTER 12

NATURE'S EDITOR

JULY 23–28, 1878—

Wyoming Territory

WHILE MARIA MITCHELL SEARCHED FOR HER WAYWARD TELE-scope parts, the other scientific parties were already rehearsing for the big event. Conducting drills was considered essential to the success of an eclipse expedition, and the Americans were wise to heed the advice of Scotland's Astronomer Royal, C. Piazzi Smyth. "So many circumstances . . . have to be noted, observed, and measured, within a few seconds," he wrote, "that it is necessary to adopt some systematic division of labour amongst a number of observers, and for each to be previously practised and expert in his particular part."

In Wyoming, practice runs were underway at the three observation posts along the line of the Union Pacific Railroad. The farthest west of the camps was at Creston, the remote rail stop in the Great Divide Basin, where William Harkness presided. One of the more senior scientists at the U.S. Naval Observatory, Harkness possessed deep experience with astronomical expeditions, having observed the total solar eclipses of 1869 in Iowa and 1870 in Sicily, and having led a U.S. government party to Tasmania for the 1874 transit of Venus. Now, in Wyoming, he was leading an expedition that comprised two Naval Observatory assistants and several civilians, including Otis H. Robinson, a mathematics professor at the

University of Rochester, and Alvan G. Clark, from Alvan Clark & Sons of Cambridge, Massachusetts, America's foremost manufacturer of optical instruments. (The Clarks made telescopes for many of the nation's top astronomers, including Maria Mitchell, and had crafted the U.S. Naval Observatory's enormous Great Equatorial.) Another observer from Cambridge, though not officially a member of Harkness's party, was also at Creston. Étienne Léopold Trouvelot, an astronomical artist who worked for a time at the Harvard College Observatory, was known for his stylized illustrations of the rings of Saturn, the cloud bands of Jupiter, the Orion Nebula. His pastels drew praise at the Centennial Exhibition—"[T]heir artistic execution is excellent," wrote James Craig Watson and the judges of Group 25—though today Trouvelot evokes scorn for a different aspect of his professional life. He dabbled in entomology and, while pursuing an ill-advised scheme to breed silkworms, kept some exotic caterpillars netted in his backyard. As he later told Alvan Clark, a gale blew the netting away, scattering the insects. Those caterpillars were gypsy moth larvae, which multiplied into a massive, hungry army that began munching its way across forests of the Northeast, eventually defoliating untold thousands of acres despite costly and futile eradication campaigns. The scale of destruction, constituting one of the most devastating insect invasions in North American history, would not be evident, however, until long after the eclipse of 1878.

At Creston, life for Trouvelot, Clark, and the others hewed to a daily routine. The scientists, who slept in the railroad postal car that had brought the telescopes from Washington, emerged one by one at 6:30 A.M. to wash up at a tin basin using water drawn from a whiskey barrel. Meanwhile, a soldier from Fort Steele cooked the morning meal. "Gentlemen, breakfast is ready," came the call at seven, triggering a rush to the mess tent for coffee, tea, cold beans, butter cakes, and hard biscuits. By seven-thirty, it was time for work.

Among the first tasks was to recheck the station's latitude and longitude. Establishing one's precise location was critical to the

U.S. NAVAL OBSERVATORY ECLIPSE CAMP,
CRESTON, WYOMING TERRITORY.

usefulness of many eclipse observations; for instance, it enabled one to compare the predicted path of the moon's shadow to the actual path, and thereby to correct calculations of the moon's orbit. Harkness took out his sextant, a handheld contrivance used for celestial navigation. It consisted of a wedge-shaped frame—contoured like a generous slice of pie—with a small telescope and a couple of mirrors attached near the pointed end, on top. Along the curved bottom— where the pie crust would be—hung a graduated arc, like a protractor. By sighting through the scope and adjusting the mirrors, Harkness could precisely measure the height of the sun in the sky. At sea, navigators did this by taking readings off the horizon. On land, at Creston, Harkness employed an *artificial horizon*, a small basin of mercury, under glass, that gravity shaped into a perfectly horizontal reflective surface. By tracking the growing height of the rising sun, Harkness was able—with the aid of tables contained in Simon Newcomb's Nautical Almanac—to reset the team's clocks according to Creston's local time. (In 1878, time zones did not yet exist, and each city ran by its own unique time, set by the sun and stars. The official time in Washington, for instance, lagged one minute, thirty-two seconds behind that of Baltimore.) Around noon Creston time, as the sun reached its zenith, Harkness took more readings with his sextant. He used them to calculate the station's latitude: 41 degrees, 43 minutes, 34 seconds north of the equator.

Finding longitude—one's distance east or west on the globe—was a different matter. It required comparing one's local time to that of a distant location whose coordinates were already established. The earth turns one degree every four minutes, so the difference in time between two places can quickly be converted into the degrees of longitude separating them. In an earlier era, this exercise would have required bringing to Creston a rugged clock that could carry the time from home without varying its rate even after weeks of travel, but by the 1870s, a much easier method could be employed. Tiny Creston had a telegraph station, and at pre-established intervals throughout the day, Harkness and his team received time signals down the wire from observatories in Utah, Pennsylvania, and Washington, D.C. With these signals, the men fixed their longitude: 2 hours, 2 minutes, 40.5 seconds west of the center of the dome at the U.S. Naval Observatory on the left bank of the Potomac.

The scientists withdrew the canvas roof from the crude observatory and rehearsed the activities they would perform during the eclipse. Alvan Clark went through the steps required for photographing the corona—inserting a plate into his camera, exposing it for a fixed period of from three to sixty seconds, then promptly removing it and inserting a new plate. Otis Robinson tested his polariscope, which divided light rays into those oscillating up-down versus left-right, potentially offering clues to whether the corona shone by its own light or merely reflected light from the sun. Harkness practiced using his spectroscope with a fluorescent eyepiece (containing uranium) that enabled him to see the ultraviolet end of the spectrum, which he hoped would reveal new information on the chemical composition of the corona and prominences. After each rehearsal, the men rehearsed again, to make sure they could perform every step efficiently in the 176 seconds that totality would last at Creston. Other than a break for lunch at one o'clock, another for supper at six, and an evening stroll, the men worked until ten. Then they reentered the postal car to doze in the cool night air, on buffalo-robe mattresses beneath army blankets, roused briefly each night by the westbound mail train that thundered past at 2:00 A.M.

———

FOURTEEN MILES TO THE EAST, a similar routine unfolded at the government eclipse camp overseen by Simon Newcomb. It too sat near a remote rail stop that had no stores, no hotel, no town—just a tiny depot and a water tank for refilling the steam locomotives— and was surrounded by "a barren wilderness of sand, where nothing grows but some low shrubbery," as Newcomb described it in a letter home. "We sink most ankle deep at every step." The wife of the stationmaster cooked meals, but the scientists here, unlike at Creston, did not enjoy the relative luxury of a railroad car to serve as a dormitory. Instead, they slept in tents, pitched in a sheltered nook beside the tracks. A sage-covered dune to the south and west helped block the prevailing winds, which often blew ferociously. Though barely a dot on a map, the rail stop had a name: Separation, because it was here that surveyors plotting the transcontinental railroad parted ways.

Still farther east, at Rawlins, Edison and his companions maintained an easier existence, living at the Railroad Hotel and working just across the tracks at the borrowed residence of Union Pacific Master Mechanic Robert Galbraith. As head of the party, Henry Draper selected the site for the observatory. He was concerned about the wind, so he had carpenter William Daley build the structure in the lee of Galbraith's house. The rough building, sixteen feet long, included a darkroom (supplied with water from a hydrant) and space for various telescopes, spectroscopes, and polariscopes. The tasimeter, given its sensitivity to extraneous sources of heat, required a separate structure. For this purpose, Edison retrofitted the Galbraiths' chicken coop.

The New York Herald's Edwin Marshall Fox followed the work closely. "The preparations for observing the forthcoming total eclipse of the sun from this point are almost completed. To-day the last nail was driven in the temporary structure of pine boards which is to serve as an observatory," he described in a dispatch on Tuesday,

July 23. "All [the instruments] are now in readiness, and several preliminary experiments have demonstrated that transportation has not injured them. The main point sought to be determined by Professor Draper is whether the corona or halo surrounding the sun's disk is only a glowing gas, or whether it contains, in addition, solid or liquid particles that reflect light from the sun to the earth." Fox wrote that the corona (actually, he meant a layer just below it, called the *chromosphere*) was thought to consist of hydrogen "together with, in a minor degree, some unknown substance, which to this day continues a mystery. This unknown material is designated by the term 'helium.'"

The following day, the Draper party took time out for an excursion. The team hopped a train to Newcomb's camp, at Separation. While the astronomers undoubtedly discussed equipment, preparations, and theories, Edison pursued other activities. He carried his new .44-caliber Winchester rifle and hoped to indulge in a little hunting. Antelope were abundant in the area, but they kept their distance and proved impossible to hit, given their alertness and speed. So Edison introduced himself to the local telegrapher, and after reminiscing about his own days as a lightning jerker and inquiring about shared acquaintances, he asked if there was other game in the vicinity, perhaps jackrabbits. "Oh, yes, plenty of them," the man said, then shaded his eyes with his hand and scanned for a long-eared hare among the brush and cacti. The animals could be difficult to spot, as their drab coloration blended with the bleak, rocky terrain. Eventually he said, "There's one, off there."

Edison unslung his rifle, carefully aimed, and squeezed the trigger. The bullet flew, but the jackrabbit remained immobile. The great inventor crept forward, reloaded, and fired again. Still, the animal sat fixed, failing even to blink. Frustrated, Edison continued to stalk his quarry. After shooting several more rounds, to suppressed laughter from onlookers at the station, he was near enough to identify the source of his trouble. The jackrabbit was inanimate, a stuffed mount that stared with glass eyes. It turned out that the local telegraph operator was also a taxidermist. He had been forewarned that Edison was on his way in search of game and placed the stuffed jackrabbit in the sage as a prank.

Edison took it in stride. He enjoyed a good practical joke—"Well, that's one on me" was one of his favorite expressions. As a scientist, however, he still aimed to be taken seriously. That evening, back in Rawlins, a man whose praise or disfavor carried enormous weight in the scientific world arrived on the midnight train.

NORMAN LOCKYER, THE BRITISH astronomer who had discovered helium in the sun, had not expected to visit America for the eclipse of 1878. Although he had led eclipse expeditions twice before (Sicily in 1870, India in 1871) and was eager to do so again, his son was gravely ill in the early part of 1878. "I have a sick boy at home otherwise I should certainly have come out," Lockyer wrote to Simon Newcomb in June, apologizing for having to miss the great event. Then Lockyer's fortunes changed, tragically. "Since my last note to you I have lost a dear child," he wrote again at the beginning of July. "[M]y

doctor's orders are presently to go away: so I have made up my mind to come and add my mite of work." With just four weeks until the eclipse—and an ocean and continent to cross beforehand—Lockyer had no time to organize a government expedition. He would come alone, as a private citizen.

Newcomb replied with condolences and an invitation for Lockyer to join him in Wyoming. "The military furnish my party with an

JOSEPH NORMAN LOCKYER.

encampment, so I have no doubt I could supply you with a tent to sleep in, if you prefer it to an open sky," he wrote. "Be sure to bring heavy blankets or rugs, or both, as the nights are extremely cold, though the days are warm. With a flask of brandy in case the water is unwholesome, I think health and comfort will then be assured."

Although Lockyer was a gregarious man who enjoyed good relations with a number of American astronomers, he did not shy away from rebuking people or ideas that he found wanting, especially when it came to his area of expertise, the sun. He had recently chided Henry Draper in the press when Draper claimed to find evidence of oxygen in the sun—evidence that would later prove faulty. Lockyer even scolded painters who, he contended, portrayed sunlight and its effects inaccurately. His comments on a new work by the English landscape artist John Wright Oakes read as follows: "Sky colours impossible with so high a sun."

Lockyer's reputation was perhaps best summarized in verse, in a poem attributed to the physicist James Clerk Maxwell:

And Lockyer, and Lockyer,
Gets cockier, and cockier;
For he thinks he's the owner
Of the solar corona.

Further amplifying Lockyer's brash personality was his professional position. He was more than an astronomer; he was the founding editor of the British journal *Nature*, one of the most influential scientific publications on both sides of the Atlantic. Through the journal, he spoke from on high, his pronouncements godlike. "Lockyer sometimes forgets that he is only the editor, not the author, of Nature," quipped a colleague.

On Thursday, July 18, Lockyer steamed into New York Harbor on the *Baltic*, his arrival heralded by the *Tribune*. It was his first trip to the United States, and he had come, the newspaper reported, not only to see the eclipse, but to observe the observers—that is, the

American astronomers, "for whose scientific attainments and methods he had the highest respect." Lockyer sent a dispatch to Newcomb saying to expect him soon.

The following Wednesday—the day of the jackrabbit incident—Lockyer crossed into Wyoming in a Pullman car on the No. 3. He received a telegram from Newcomb: "[S]ome of us will meet you at Rawlins." When the train arrived, Newcomb was there. The two men had met eight years earlier, when Newcomb passed through London, with his wife, on the way to Gibraltar for the eclipse of 1870. The Lockyers had been gracious to their American guests. "There was quite a party assembled to meet me," Mrs. Newcomb wrote after an evening at the Lockyer home. "Refreshments were constantly passed around." Now, in the United States, Newcomb was the host. He escorted Lockyer to Separation. Apparently there was no tent for the Englishman, however, as he spent the night in the tiny depot. The next day, Lockyer returned to Rawlins and checked in at the Railroad Hotel.

MEANWHILE, ANOTHER SELF-IMPORTANT astronomer secured a room at the hotel. Planet hunter James Craig Watson had left Ann Arbor Monday evening and arrived in Rawlins late Thursday, carrying with him two items of note. One was a telescope—a new refractor that he had borrowed from Michigan State Normal School (later Eastern Michigan University) because his own university did not possess an instrument that could be conveniently transported. The loaned telescope was a handsome piece of craftsmanship. Its gleaming brass tube stretched more than five feet from eyepiece to objective lens, the latter four inches in diameter. The equatorial mount allowed the telescope to pivot easily atop a mahogany tripod, which stood at shoulder height. Etched into the faceplate, in elegant cursive, was the name of the manufacturer: Alvan Clark & Sons.

Watson also brought his wife. Despite being well educated and slightly older than her husband, Annette Watson seemed resigned to the role of subordinate spouse. "[She] is in every thought devoted to her liege lord," wrote one who encountered the pair during their time in the West, although another astronomer, the affable Princeton professor Charles Young—who had accompanied the Watsons to China for the 1874 transit of Venus—was far from good-humored in describing the couple's relationship: "[H]is treatment of his wife was simply abominable. She was rather weak & querulous, but gave no reason for his abuse, which sometimes almost went to physical violence." The two had no children, yet there was some evidence of love between them. Watson honored his wife in the heavens by christening one of his asteroids *Helena*—Annette's middle name. On earth, he invited her not only to accompany him to Peking for the transit of Venus, but afterwards to travel together on a long, arduous, and sometimes romantic trip home, visiting the Taj Mahal by moonlight, crawling inside the Great Pyramid of Giza, climbing Mt. Vesuvius, and riding gondolas on Venice's Grand Canal. "I do enjoy traveling & sightseeing—am never weary of it," Annette wrote her parents from Egypt. She did occasionally weary of her husband, however, as she noted from Hong Kong: "J. C. is so nervous & almost sick from a cold taken in Shanghai that I can scarcely endure his impatience." More than impatient, he was also loath to reveal his own frailty. "He has just told me not to write that he is sick, but it is already done," Annette added with evident exasperation. "I never heard such a cough in all my life. He will never hear to me, but thinks he knows better how to take care of himself."

By Saturday—two days before the eclipse—the tiny Railroad Hotel was overflowing with guests. Daniel Hector Talbot, a prosperous land broker from Sioux City, Iowa, stopped over in Rawlins while on his way to the Pacific Northwest for business. Talbot was passionate about science and would later spend his fortune collecting natural history specimens, then donate thousands of bird skins to the University of Iowa before dying destitute. W. Fraser Rae, a British journalist and author, had come over from London to accompany Norman Lockyer. Meanwhile, two much younger

Englishmen—recent graduates of Cambridge University—had also sailed on Lockyer's ship across the Atlantic and arranged to meet up with him for the eclipse. "At a little after midnight arrived Rawlins. No beds," wrote one of the pair, R. C. Lehmann, a handsome fellow with a square jaw and dimpled chin who would later become a well-known magazine writer and parodist. "Roused up Rae & Lockyer. I slept on floor at foot of Lockyer's bed."

Rawlins had by now become a veritable Athens of the West. Such a collection of great minds, in such an unlikely setting, was an event worth preserving for posterity, and it so happened that a professional was on hand to do just that. J. B. Silvis was a former saloonkeeper and failed gold miner who found success as a roving photographer on the rails of the Union Pacific. "[He] meanders up and down the U. P. in his palatial photograph car, seeking the shadows of us poor

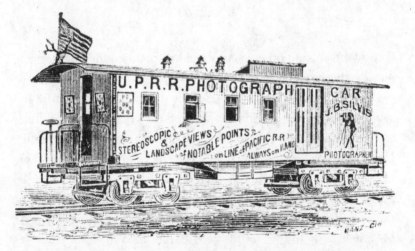

terrestrial mortals," a Nebraska newspaper explained. The "palatial" car was a converted caboose, outfitted on the inside with a darkroom and portrait studio, and ornamented on the outside with advertisements for his business and an American flag that flew above a severed, antlered elk's head.

The Draper eclipse party, along with Norman Lockyer, James

THE ECLIPSE STATION AT RAWLINS, WYOMING TERRITORY.

Left to right: George Barker, Robert Galbraith, Henry Morton, unknown, Fred Hess,
D. H. Talbot, W. Fraser Rae, Edwin Marshall Fox, James Craig Watson, Annette Watson,
Anna Palmer Draper, Henry Draper, Thomas A. Edison, J. Norman Lockyer.
From a photograph by J. B. Silvis.

Craig Watson, and other visitors and locals, posed for a photograph outside the crude Rawlins observatory on Saturday, July 27. The twelve men and two women stood along the wall of the structure and against the picket fence that surrounded the Galbraiths' yard. Telescopes could be seen behind them, emerging through the top of the Draper observatory and the side of the chicken coop, and two more telescopes stood in the dirt before them. Most everyone was adorned appropriately for a portrait, the women in subdued Victorian finery, the men in dark suits and bow ties. Watson looked especially dapper, his ample belly filling out his vest. Barker held his head erect, his neatly combed hair in contrast to his signature shaggy beard. Then there was Edison. He crossed his arms before his boyish frame. His bangs hung unevenly, not yet grown out after the recent encounter with wet plaster. His clothes, to put it generously, were casual. "He was the worst dressed man in the room," a reporter noted when Edison wore much the same outfit two weeks later. "An old black hat, a cheap shirt with the stud-holes in the bosom unoccupied, a two-bit necktie several months old, coarse pants and vests and a mouse-colored linen duster, completed his attire."

At night, this eminent assemblage invited the people of Rawlins to look through their telescopes for tours of the night sky. The townsfolk gazed at double stars, distant nebulae, and Saturn's rings, which at the time were angled directly toward Earth and gave the appearance of a knife edge slicing through the planet. The heavens shone so clearly that the sharp-eyed Watson claimed he could spot the moons of Jupiter without a telescope. "[The scientists] never tired of showing and explaining to the citizens the use of the instruments, and showing them the wonders of the heavens through their glasses," wrote the *Laramie Daily Sentinel*. "Such an array of distinguished scholars, so provided with the most improved fixtures, furnished a rare opportunity to us frontier residents to enjoy some of the wonders of science."

It was also a rare opportunity for the astronomers themselves—to reconnect with old colleagues, who were also competitors. The tension could make things a bit awkward, as Lehmann, the young Cambridge man, sensed. "[Draper's] great reputation rests on the discovery of oxygen in the sun; Lockyer has written strongly in refutation of this. They dine at the same table here & preserve a sort of armed neutrality." Later, as the scientists sat in the Draper party's rustic observatory, they all reminisced, boasting of past expeditions. Norman Lockyer recalled his last eclipse, in India, where he slept in a hammock, in a jungle with scorpions and snakes, while doped up on opium to treat a fever. His party made its base an abandoned fort on the coast, and when it erected its instruments, the locals mistook the telescopes for guns and braced for war. The astronomers successfully calmed those fears, but the eclipse provoked new ones. Lockyer recalled the wails and lamentations of thousands who surrounded the fort and—as he described it—begged the scientists to rid the world of the dreadful dragon that had swallowed the sun.

James Craig Watson shared his own tale of peril and superstition, from his time in Peking for the transit of Venus. In Chinese astrology, the sun represented the emperor, and therefore a planet's black silhouette crossing the sun portended ill. Just as the celestial event came to pass, the omen appeared validated when the young Emperor Tongzhi fell sick, diagnosed with smallpox (although some suspected the sexually adventurous teen had really contracted syphilis), and he soon expired. The Chinese public blamed the visiting astronomers, and it was only by good luck and strategy, Watson said, that he escaped China alive.

Watson likely told another story from China, a favorite of his, about his discovery of an asteroid by accident. It was the night of October 10, 1874—two months before the transit of Venus—when, while setting up his post in Peking, he turned his telescope toward Pisces. There, among a sea of stars, he noticed one shining dimly that seemed out of place, based not on a star map but simply on his memory of that patch of the heavens. Subsequent observations showed that it was a new minor planet. "This being the first planet discovered in China, I requested Prince Kung, regent of the Empire, to give to it a suitable name," Watson would write. The asteroid became known as Juewa—"China's Fortune"—but it was really Watson's good fortune. He had shown his talents to be almost superhuman. The man could glance at the sky and pluck out a planet. What chance, then, did Vulcan have to hide during the eclipse?

THE WEEKEND BEFORE NATURE'S big show was devoted to final preparations. For Watson, that meant jury-rigging his telescope. Although it was a fine instrument, it lacked setting circles—graduated disks that indicate where the telescope is pointed, measured in *declination* and *right ascension* (the celestial counterparts to latitude and longitude). Without these coordinates, Watson would have difficulty specifying the location of any planet he might find, undermining his ability to substantiate the discovery. So he improvised a solution, akin to making his own setting circles. He hired a Rawlins carpenter to fashion two wooden disks, onto which he pasted white cardboard. These he attached to the telescope along with brass pointers that moved just above the cardboard to indicate the instrument's alignment. The plan for the eclipse, then, was to mark in pencil, on the white circles, the position of the brass pointers when the telescope was aimed at a reference object—say, the sun—and then again when it was aimed at an unknown object of interest. Later, after the hurried work of totality, he would dismount the wooden circles and, at leisure, measure the angles between the marks to come up with the celestial coordinates. It was, he was convinced, a foolproof technique. It would permit no error—and no doubt—as to where he observed an object in the sky.

Edison, too, needed to prepare his equipment. The tasimeter was not yet ready for the eclipse. Just hooking up the device to its external battery and galvanometer (for measuring electric current that indicated heat) proved enormously complicated; he had to conceive a whole new way of wiring the contraption to get it to work. Next, there were continuing problems caused by the device's sensitivity to extraneous temperature changes. "The approach of any person within five feet threw the instrument out of adjustment," Edwin Marshall Fox reported in the *Herald*. "The heat from his little finger at that distance deflected it several degrees." So Edison insulated the tasimeter; he placed it in a double tin case that held water in a kind of moat around the device. This ungainly assemblage then had to be attached to a telescope—a four-inch refractor that George Barker had brought for Edison's use—and placed within the confines of the Galbraiths' chicken coop.

Still, Edison had yet to demonstrate that his apparatus could do what he said it could do: measure the heat of a star millions of miles away. So, a few nights before the eclipse, he performed a test on Arcturus, the brightest star in the northern half of the sky. He aimed the telescope high overhead and carefully focused the starlight onto the rubber strip inside the tasimeter, which rested on a stable platform above the earthen floor of the hennery. Edison was on his hands and knees. The other scientists gathered around and watched.

Edison carefully eyed the galvanometer's needle, which pointed to zero—no indication of heat. Minutes passed. Still, the needle did not budge. "It's strange," Edison mumbled. "It ought to work, and I'm sure it will." He examined the connections and discovered the problem; a wire was out of place. He reconnected it.

Starting over, he again aimed the starlight into the tasimeter. This time the needle reacted immediately, deflecting toward the side of heat. The instrument was so delicate, however, that it sometimes gave spurious results. He needed to be sure that this one was real. So he placed a dark screen in front of the telescope to block the starlight. The indicator returned to zero. Then he removed the screen, allowing the light through again. The needle jumped, registering heat. He repeated the experiment several more times until he was convinced that the findings were valid. His tasimeter really could detect heat from Arcturus. Edison then turned his telescope to another star and repeated the exercise. He continued his labors until 4:00 A.M.

"One of the many points of interest here, to me, has been the observatory in which Mr. Edison has been experimenting on his tasimeter," Norman Lockyer wrote to London, in a dispatch for *Nature*, the weekend before the eclipse. "It is truly a very wonderful instrument, and from the observations made last night on the heat of Arcturus, it is quite possible that he may succeed in his expectations. For its extreme delicacy I can personally vouch. The instrument, however, is so young, that doubtless there are many pitfalls to be discovered. Mr. Edison, however, is no unwary experimenter."

Lockyer, the grandiloquent European, cast his praise wider still. Having watched a number of U.S. astronomers prepare for the eclipse, and having heard of the work of others, he lavishly commended the scientific preparations of the neophyte nation. "The energy displayed by the American astronomers is, if possible, greater than I anticipated," Lockyer wrote. "There is scarcely a man of note among them who is not now along the totality line which runs from the Yellowstone Park to the Gulf of Mexico."

Although the astronomers appeared ready, Lockyer mentioned one concern: the weather. The skies in recent days had clouded up each afternoon, a worrisome pattern that, if it continued on Monday, could negate all the hard work in preparing for the eclipse. And it was not just a problem in Wyoming. "[A]t Denver," he added, "matters have been much worse."

CHAPTER 13

OLD PROBABILITIES

JULY 28, 1878—

Colorado

"RAIN?—OH, NO, IT DOESN'T 'RAIN' IN COLORADO THIS YEAR— it just lets go all bolts and comes down in sheets, in torrents," a newspaper from a mountain town near Denver griped the weekend before the eclipse. "During the past eight years, Colorado has not witnessed a season marked by so many heavy rainfalls—they call 'em 'cloud bursts,' and the effects justify such nomenclature." Day after day, it was the same. Even if morning dawned to clear skies, afternoon brought downpours at the very hour when the eclipse was set to occur.

On Friday, July 26, Maria Mitchell, after much duress, finally retrieved her lost luggage from the Denver & Rio Grande Railway, but the weekend storms made it impossible to set up her telescopes and rehearse. (The tempests likely frayed her nerves, too, for she suffered from a lifelong fear of lightning—ironic for a cousin of Benjamin Franklin.) At the Princeton camp on Cherry Creek, the water rose so high one night that it almost swamped the tents. On Sunday morning, at area churches, there was just one thing to do—pray— "which is precisely what every one who has faith in the efficacy of prayer and a proper regard for science should do," wrote the *Denver Daily Times.* Still, Sunday afternoon brought even more rain, and then hail. The cumulus clouds "hovered over our fair city like birds of ill omen," as one scientist put it, "reducing to zero the hopes of astronomers, rousing the ire of many of our citizens and exciting the anxieties of all." But of all the scientists bemoaning the inclement weather on that day before the great eclipse, the one who agonized the most—whose life was actually endangered by the atmosphere— was the man who had predicted clear skies.

METEOROLOGY REMAINED AN IMMATURE and imperfect science in 1878. During a recent speech in New York, Mark Twain had lampooned those who tried to prognosticate the weather. "Probable northeast to southwest winds, varying to the southward and westward and eastward and points between," began his parody of a forecast, which continued with "probable areas of rain, snow, hail, and drought, succeeded or preceded by earthquakes, with thunder and lightning." Twain's hapless weatherman then concluded: "But it is possible that the programme may be wholly changed in the meantime."

The critique was not quite fair. Scientists had made great strides in understanding how weather systems formed and moved. Indeed, a key insight had come more than a century earlier, thanks to an eclipse that was *not* successfully observed. In the fall of 1743, Benjamin Franklin planned to view a lunar eclipse one evening in Philadelphia, but as the appointed hour approached, a violent storm did as well, "so that neither Moon nor Stars could be seen." It was a northeaster— so named because the winds blew from that direction. "The Storm did a great deal of Damage all along the Coast," Franklin wrote, "for we had Accounts of it in the News Papers from Boston, Newport, New York, Maryland and Virginia. But what surpriz'd me, was to find in the Boston Newspapers an Account of an Observation of that Eclipse made there: For I thought, as the Storm came from the N E. it must have begun sooner at Boston than with us, and consequently have prevented such Observation."

At the time, it was believed that storms moved in the direction of their winds, but this northeaster had moved against its winds, *toward* the northeast. Further study convinced Franklin that such storms along the East Coast generally traveled in that same direction— from Georgia to Nova Scotia. This finding carried practical implications. If a storm set in and you knew its direction of movement, you could give downstream cities an early warning of its impending arrival. At least, you could do so if you had a way to relay the message fast enough.

The telegraph made such real-time storm warnings possible. The Smithsonian's Joseph Henry, whose electrical research had paved the way for Morse's invention, took an initial step toward that goal in the 1850s. He arranged for an assemblage of telegraph stations each morning to wire local weather conditions to Washington, where the information was posted at the Smithsonian (using colored cards on a large map) and published in the newspaper. The continually updated chart gave a general sense of storm systems moving across the country, but Henry's institution did not issue formal forecasts. The federal government did not take on that responsibility until a decade later, when a nascent weather service rose from the ashes of the Civil War.

During the war, Union commanders had relied heavily on military intelligence. They gathered information on Confederate troop movements from spies and scouts, and from a specially trained team—the Army Signal Corps—whose men climbed high hills and church steeples to scan the terrain and send back coded messages using flags, torches, and telegraphs. When the war ended, such skills were no longer in demand, but the Signal Corps' leader, General Albert J. Myer, was an astute political operator. With his team's budget slashed and its very existence threatened, he sought a new peacetime role. He found it in the late 1860s, when a spate of shipping disasters on the Great Lakes prompted a Wisconsin congressman to call for the establishment of a federal storm-warning system. Myer argued that his corps possessed just the skills required:

it could gather meteorological data at military posts across the country and then telegraph those statistics back to a central clearinghouse, "giving the presence, the course, and the extent of storms . . . as it would, in time of war, those of an enemy." His lobbying succeeded. On February 9, 1870, President Ulysses S. Grant signed legislation creating America's first national weather service. It would be operated by the Army Signal Corps.

GEN. ALBERT J. MYER.

General Myer had secured a fresh mandate and a vastly increased budget. What he lacked was a meteorologist.

CLEVELAND ABBE HAD NOT started his career in meteorology. His first love was astronomy, an odd passion for a severely nearsighted boy. ("[W]hen mother had taken me on her lap singing 'Twinkle, twinkle little star,' I had looked up at the sky and wondered but saw nothing," he recalled.) Fitted with strong spectacles and using a borrowed telescope, he studied the night sky from the roof of his family's New York City home. After college, he did advanced work in astronomy at the University of Michigan—where he came to

CLEVELAND ABBE,
"PROBABILITIES."

know James Craig Watson—then served as a kind of postgraduate fellow in Russia, at the well-appointed imperial observatory near the tsar's summer palace south of St. Petersburg. But what Abbe really wanted was a permanent, paying job in astronomy, a search that proved an exasperating struggle for more than a decade. He applied for the new professorship in astronomy at Vassar, but he lost out to Maria Mitchell. Like Mitchell, he worked for a time as a government computer, but he nearly collapsed under the stress of numerical calculations. ("It was physical-, mental- and almost moral-suicide," he confessed to Simon Newcomb.) Abbe even tried to marry into the profession, it seems, offering his hand to a younger sister of the man who oversaw Russia's imperial observatory. She declined.

Finally, in 1868, Abbe landed what appeared to be—on paper, at least—the ideal job: director of the storied Cincinnati Observatory. The institution had emerged a quarter century earlier out of a grassroots fundraising campaign, pitched as democratic America's challenge to the tsar's royal observatory. "I am determined to show the autocrat of all the Russias," said the Cincinnati Observatory's founding director, "that an obscure individual in this wilderness city in a republican country can raise here more money by voluntary gift in behalf of science than his majesty can raise in the same way throughout his whole dominions." The good citizens of Cincinnati answered the call, donating enough money to buy one of the world's most powerful telescopes, and in 1843 the cornerstone was laid by an elderly John Quincy Adams, who had marveled at the 1806 total solar eclipse while a U.S. senator and, later, during and after his presidency, had advocated the construction of astronomical observatories. (He called them "lighthouses of the skies.") But success for the Cincinnati Observatory proved short-lived. By 1859, its director left, and its funds dwindled. When Cleveland Abbe arrived a decade later, the building was all but abandoned, its windows broken and its roof leaking. Out on the portico, one of the Greek columns was tipping over.

Abbe began repairs to the edifice, but he could not fix a bigger problem: the area's worsening air pollution. Cincinnati—dubbed Porkopolis—was a booming city of meatpackers and brewers, and thick smoke often blanketed the sky. Since atmospheric conditions dictated how much one might see through a telescope, Abbe grew interested in meteorology, and he devised a plan to track the weather in Southwest Ohio. With the help of observers who telegraphed data from a broad geographic region, Abbe drew up daily forecasts—he called them *probabilities*—and, beginning in September 1869, he offered the service to area newspapers and private subscribers. For his efforts, this fledgling weatherman, though thirty years old, earned a fondly geriatric nickname: "Old Probabilities," or "Old Probs."

Cleveland Abbe had created something historic—the first regular weather forecasting service in the United States—but within a year the chronically cash-strapped Cincinnati Observatory was forced to abandon the venture, and Abbe too was soon dispatched, placed on unpaid leave. General Myer, just then seeking to create a weather service on a national level, called on Abbe for advice and quickly offered him a job as his chief meteorologist.

At last finding stable employment, Abbe moved to Washington. Recently married, he soon fathered a son, then two more. From his respectable salary he bought a sizable townhouse on I Street, with a lunette-topped doorway in front, a garden in the rear, and, inside, plaster cornices adorning twelve-foot ceilings. It had once been home to James Monroe and, at the beginning of Monroe's presidency, had served as the executive mansion while the White House, burned by the British in the War of 1812, underwent final repairs.

Abbe's office was a short walk away, on G Street, where the Army Signal Service (as the Signal Corps had now come to be known) occupied a three-story brick building topped by weather vanes, rain gauges, wind meters, and other "toys which excite the envy of all the neighboring boys," as one observer put it. A tangle of telegraph wires led inside, funneling in weather reports taken simultaneously at meteorological stations across the United States. The messages were condensed using a cipher, to speed transmission and reduce cost. *Ransack* meant that 0.49 inches of rain had fallen since the last report. *Dagger* was a barometric reading of 29.63. The word *cake* indicated fair skies, about half covered by clouds, with winds from the south. The reports arrived thrice daily, setting off a frenzy of work among a half dozen clerks. Standing at desks, they mapped precipitation, clouds, and wind velocity; drew isobars and isotherms to reveal patterns of air pressure and temperature; and then Abbe or one of the forecasters he trained would deduce how the weather was

THE SIGNAL OFFICE AT WASHINGTON.

likely to change over the next day or two. Within an average of one hour and forty minutes, the latest forecast was ready—telegraphed to the press for publication nationwide.

Cleveland Abbe's new job earned him a comfortable life and the respect of the scientific community, but he bristled under the office rules. He was a civilian of academic bent stuck in a military hierarchy. When George Barker invited him to address a scientific gathering in Philadelphia in 1876, Abbe drafted a reply explaining that he was forbidden to say or publish "anything relating in the remotest degree to our work without first submitting it to General Myer (and whatever I submit to him is tabled)." He wrote further, "Lectures or any work by means of which the individuals of the corps may build up for themselves an independent reputation are especially obnoxious." Before sending the final letter, Abbe diplomatically removed any direct mention of his boss, but it was obvious that he had interpersonal differences with General Myer. They were very different men.

Whereas Cleveland Abbe was gentle and bookish (his favorite pastimes were long walks and croquet), General Myer was rigid and domineering. While Abbe argued that the Signal Service must do more than just churn out forecasts—it should conduct basic research to advance the field of meteorology—Myer evinced little interest in theoretical studies. And there was another matter that, for Abbe, must have chafed. His superior had usurped his nickname. The newspapers had taken to calling General Myer "Old Probabilities."

One summer evening in 1874, while Cleveland Abbe was at home reliving the bachelor's life (his wife and young sons were away on a seaside vacation), General Myer stopped by. When a servant informed the visitor that the man of the house had already gone to bed, the general laughed—"What, so early?"—and left. The next day, Abbe recounted the incident in a letter to his wife. "Somehow I always feel better when I thus miss seeing him," Abbe admitted. "I suppose I am foolish & wicked but I can't help pining for freedom & a telescope."

IN EARLY 1878, as astronomers began planning for the coming eclipse, they called on the Signal Service for an obvious reason. "I shall be greatly obliged if you will have the goodness to furnish the Observatory, if possible, any data from your records that will guide us in selecting [viewing posts] such as may be thought to give greatest promise of good weather," Admiral John Rodgers, the head of the Naval Observatory, wrote to General Myer. Myer handed the request to Cleveland Abbe.

What was required was not, strictly speaking, weather forecasting; there was no way to predict the movement of high- and low-pressure systems several months in advance. The task was simply one of compiling and analyzing historical data. Abbe identified thirty-six weather stations that fell within the computed path of the moon's shadow, and another thirty-one that sat just outside. He gathered statistics for the months of July and August over the previous years, and he then calculated the odds of clear skies on the afternoon of July 29. According to Abbe's math, the region around Rawlins, Wyoming, should enjoy a 79 percent chance of favorable conditions at the time of the eclipse. At Denver, the numbers were less optimistic, but he still calculated better than even odds of a clear view, at 60 percent. The Signal Service issued these figures, and many others, in a circular that it distributed to the government and the press.

But Abbe, the frustrated astronomer, aspired to do more than provide advice; he hoped to observe the eclipse himself. He had experienced totality once before. On August 7, 1869—when Maria Mitchell, James Craig Watson, and Simon Newcomb stood within the moon's shadow in Iowa—he was in Dakota Territory, where he had led a wagon expedition across the prairie toward the northwestern end of the shadow's path in the United States. For three precious minutes, he studied the corona and stood hypnotized, dazzled by twisting cones of light that rose like pearly mountains from the sun's hidden disk. The spectacle ended too soon. "The totality had passed away like a dream," he wrote, "and no earthly power could recall that shaded sun to allow us only a moment's longer study of its surface."

On that same day in 1869, General Myer had been near the other end of the path of totality, in southern Virginia. In an effort to view the eclipse above the smoke and haze of the lower atmosphere, he had climbed a five-thousand-foot peak. "[O]ur party was one of the very few, possibly the only one, that observed successfully from a mountain top so elevated," he wrote. What he saw melted the heart of this hardened military man. It was, he recalled, "a vision magnificent beyond description."

Therefore, when Myer learned where the 1878 eclipse would cross Colorado, he immediately focused on a prominent geographic feature that sat in the middle of the path. This was Pikes Peak, which rose more than fourteen thousand feet above sea level, dominated the horizon south of Denver, and held on its top a Signal Service meteorological post, the highest-altitude weather station on the planet. Manned year-round and connected to the outside world by a seventeen-mile telegraph line, the post imposed harsh duty on its small staff of resident soldiers, who endured electrical storms, hurricane-force winds, frostbite, and the nausea, headaches, and dizziness brought on by the thin atmosphere. Myer was intrigued. Having witnessed the 1869 eclipse from a mountaintop in Virginia, he was curious to know how this eclipse would appear from an even loftier height in Colorado. He hoped to go to Pikes Peak himself, but even if he could not, he wanted another qualified observer on the summit. After all, it was only appropriate for the nation's meteorological bureau to take advantage of the eclipse to study the sun, since the sun drives the weather on earth.

On this rare occasion, General Myer and his chief scientist saw eye to eye. They agreed: Cleveland Abbe should go to Colorado. He should climb the mountain to observe the hidden sun.

ON FRIDAY, JULY 20, Cleveland Abbe arrived in Colorado Springs, near the base of Pikes Peak, and checked into the Crawford House. The hotel, at two dollars a night, did not especially impress him (another guest around this time complained of large holes in the bedspreads and carpeting), but he was pleasantly surprised by the city—its irrigated yards, its energetic spirit. As for the weather, it could hardly have been better. "This morning the sky is everywhere as clear as a bell and the blue is such a deep fine solid blue that it's inspiring to an astronomer to look at it," he wrote to his wife on Saturday. Less fine were his preparations to ascend the mountain. "There's a good deal of confusion," he acknowledged.

Abbe was not to be the sole astronomer on the summit. Samuel Pierpont Langley, who had encouraged Edison to invent the tasimeter (but who never did receive one before leaving Pittsburgh), knew

COLORADO SPRINGS.

Abbe from the days when they both toiled as novice directors at poorly funded observatories along the Ohio River. "I am very glad to hear from you, and to exchange experiences with any one interested in Astronomy," Langley, in Allegheny, had written to Abbe, in Cincinnati, in 1868. "[H]ere, there is no one to talk to." Now, in 1878, the two men arranged to join forces on Pikes Peak. "It's first-rate;— your going;—we'll have a time—hurroo!" Langley wrote to Abbe the month before the eclipse, then received a kind offer of logistical help from the Signal Service. "[General Myer] takes pleasure in offering you the hospitalities of the Signal Station at the summit for such time as you may need," read the letter from Washington dated June 13.

Several weeks later, however, when Langley reached Colorado, he found the local Signal Service officers unprepared for his arrival. He had brought along, as an assistant, his brother—John W. Langley, a chemistry professor at the University of Michigan—but was told "there was no room for [either of] us in the Signal Service station on the summit." Samuel Langley had also brought twelve hundred pounds of equipment, and—as he wrote in frustration—"The distance to the Peak was, as we now learned, 18 miles by a foot-path. . . . No wheeled conveyance was possible." This was not the situation he had anticipated, but it was too late to relocate his observation post.

So an arrangement was made to hire burros, a cantankerous yet reliable mode of transport. The Langley brothers repacked their equipment into smaller boxes, which were then suspended between pairs of donkeys and ferried to the peak by drivers who hurled rocks and profanities at the animals to keep them plodding up the narrow trail. The rugged path crossed a raging mountain stream, angled up a deep gorge, passed waterfalls and crags and snowfields, and finally emerged above the timberline to reveal vast meadows of

SIGNAL STATION ON THE SUMMIT OF PIKE'S PEAK.

alpine wildflowers. At the summit, the Langleys—soon joined by Cleveland Abbe—found the weather station to be a small stone hut, overcrowded and half in ruins. Surrounding it was a broad expanse of jagged boulders. Abbe had brought tents, but there seemed to be no suitable spot to pitch them, so the men constructed platforms out of firewood. They covered the wood with damp hay, then tied the canvas to the rocks with wire.

The ensuing days brought occasional moments of pleasure. The three scientists gazed down and east across the plains, which stretched like an ocean to a shoreless horizon. They looked west over forested valleys, shining lakes, and snowcapped peaks toward a region that was still home—although not for much longer—to the Ute tribe. (The white man's encroachment was already inscribed in

"SPECTRES."

the recently named summits: Mounts Harvard, Yale, Princeton, Lincoln.) The scientists marveled at the play of light and shadow on the clouds that rolled by. But the remainder of the time—almost all of it, in fact—was torment.

The weather, as Abbe so delicately phrased it, "was more than usually unpropitious." He telegraphed Washington with updates each evening at nine. On Wednesday, he wired: "Cold rain all this afternoon." On Thursday: "Rather disagreeable weather." Friday: "Occasional glimpse of the sun, but mostly snow, fog, and clouds." And Saturday, just two days before the eclipse: "Two snow storms with gales endangering tents and instruments. No satisfactory glimpse of sun today. Very little progress in preparing for Monday."

The telegrams did not adequately convey the misery. At night, under damp blankets, the men struggled to sleep while the wind roared and shoved—intending, it seemed, to blow them, tents and all, off the mountain. In the day, during brief intervals of sunshine, the astronomers strove to set up their telescopes, but as soon as they unwrapped the equipment from its protective canvas, the clouds again unleashed rain or hail. (To prevent his telescope from rusting, Samuel Langley covered its steel parts with lard.) A reporter who scaled the mountain to interview the scientists declared them "the bluest party of astronomers that could have been found in the great state of Colorado. They had almost abandoned the hope of a favorable view of the eclipse, and were in a state of profound and unanimous disgust."

The worst of their suffering was not emotional, but physical. In the thin atmosphere, their hearts raced, and the slightest exertion left them gasping for air. "I lay awake the second night, drawing long breaths, in the vain attempt to breathe once satisfactorily," Samuel Langley wrote, "and remember thinking occasionally of a mouse I once saw experimented on, under the bell of an air pump, with a sympathy, born of new experience." The men were being starved of oxygen. They developed acute mountain sickness. "[W]e felt constant and severe headache, and nearly every symptom which attends sea-sickness," Langley recalled.

After a few days, the Langley brothers gradually began to acclimate, but not so Cleveland Abbe. His symptoms steadily worsened until, on Sunday morning—the day before the eclipse, when churchgoers in Denver prayed for clear skies—he awoke in his tent to paroxysms of pain in his head and upper back, and he was unable to rise. Abbe's inability to stand suggests that he had developed a dangerous condition known today as high-altitude cerebral edema. His brain was swelling, squeezing against the skull. The resulting pressure can rapidly cause confusion, hallucinations, coma, and—especially if one spends another night at high altitude—death.

The underlying physiology of Abbe's condition was not understood in 1878, but John W. Langley had been trained as a physician, and he could tell that Abbe's infirmity was potentially lethal. Langley implored Abbe to evacuate to a lower altitude, but Abbe refused, adamant that he should remain on the peak for the eclipse.

His stubbornness was understandable given that he had come so far, but he likely too was suffering from impaired judgment, a common symptom of high-altitude cerebral edema. So Abbe remained in his tent, under the illusion that with rest he would recuperate.

MEANWHILE, SOME EIGHT THOUSAND feet below, beginning his own ascent of the mountain, was the man Abbe preferred to avoid when possible. General Myer, "otherwise known as 'Old Probabilities,'" as a Denver newspaper wrote, had arrived in Colorado the previous day. At 5:00 P.M., the general emerged on the summit and found his chief meteorologist an invalid. Whereas the Langleys could only try to persuade Abbe to evacuate, Myer had the authority to issue a command, which he did. Abbe was thus loaded on a stretcher. By 6:30, four men were carrying him like pallbearers down the rocky, zigzag trail.

Flat on his back on the litter, Cleveland Abbe faced skyward through woozy consciousness. As the sun set for its final time before the great eclipse, he watched the heavens darken. Although in his hasty retreat from the summit he had forgotten his strong spectacles for stargazing, he could still, with his weaker prescription, pick out the constellations: Cygnus, Lyra, Aquila, Cassiopeia. The Milky Way, a vivid band across the firmament, shone auspiciously—evidence, at last, that the skies had cleared.

PART FOUR

1878

CHAPTER 14

FAVORED MORTALS

MONDAY, JULY 29, 1878

MORNING THROUGH MID-AFTERNOON

ACROSS THE BREADTH OF THE NATION, ON THE MORNING OF the great eclipse, it seemed as if a long-awaited tournament—or battle—was set to commence. New York's newspapers exuded anticipation. "[I]t will probably be the most interesting and important total eclipse ever seen by man," *The Daily Graphic* rhapsodized. *The New York Herald* explained that scientists would investigate "in a manner never before possible the theories of solar physics." The front page of *The Sun* offered the headline THIS AFTERNOON'S ECLIPSE, with the subhead: "Prof. Edison and Other Savants Ready to Watch the Moon's Passage."

A rundown of those savants appeared in *The Philadelphia Inquirer.* "Professors Newcombe and Harkness take charge of the stations at Creston, Wyoming" began the list, which, despite small errors of spelling and location, conveyed a good sense of the field of play. "Professor Langley, with General Myer and Professor Abbe, of the Signal Service, are at Pike's Peak, and various other points in Colorado are occupied. With these astronomers there are many amateur scientists, and others will make observations independent of the government programme. Professor Young is at Denver, Professor Draper at Rawlings, and Miss Maria Mitchell near by."

As to the scientific goals for the eclipse, *The Chicago Times* outlined the most important. "First, the establishment of a relative co-ordinate of the sun and moon"—that is, determining the precise start and end times of the eclipse at different locations, which would enable the Nautical Almanac to update its tables of the moon's orbit. "Second, the study of the physical constitution of the sun by an examination of the corona and protuberances that jut out from behind the moon when the sun's disc is wholly obscured." In this regard, Edison's tasimeter was a new tool that could offer new insights. "A third matter of interest," the paper

PATH OF THE ECLIPSE

continued, "is the opportunity the total eclipse affords in searching for any planetoid or group of planetoids that may be between Mercury and the sun"—in other words, Vulcan. *The Washington Post* left no doubt that this last trophy was the most coveted. "Should this body be discovered, it would be one of the greatest triumphs that astronomy could achieve."

The Boston Globe ended its preview of the day's event on a patriotic, self-congratulatory note, reminding its readers that eclipses were once seen as omens that portended "accident, the coming of disasters, and tokens of the anger and wrath of the Creator." Not so in modern, enlightened America. "Science and general education," the paper asserted, "have banished all the dread which these events inspired."

THERE WAS AMPLE DREAD, though, among the scientists at their camps in Wyoming and Colorado. The depths of anxiety experienced by an astronomer in the hours before a total solar eclipse are difficult to fathom. With so much to do and so much to go wrong, emotions can overwhelm. One British scientist who headed an eclipse expedition to Siam in 1875 recalled that, the day before the event, "I could not help sitting down and having a good cry."

At Creston, William Harkness and his party emerged from their postal-car sleeping quarters to a chilly sunrise and nervously eyed the heavens over the Great Divide Basin. "[N]ot a cloud was to be seen in the deep-blue sky stretching above us in all its purity," wrote an enthusiastic E. L. Trouvelot. Harkness too was optimistic. "Everything promised well for the eclipse," he remarked. The men washed up, then sat down for breakfast. A wind blew in from the southwest. It quickly strengthened, propelling dirt airborne. By eight o'clock, the astronomers in the mess tent found themselves and their dishes covered with sand and dust.

Down the tracks in Rawlins, the Draper party scanned the skies. They anxiously watched a cloud bank thicken in the east, but a few hours later—to their relief—it moved off toward the south. By noon, however, the wind picked up here, too, rocking their frail observatory. Even more vulnerable to the gusts was the chicken coop that housed the tasimeter. Edison had spent the weekend carefully adjusting his instrument, but the gale was now undoing his hard work— throwing the equipment out of alignment. Frantic, Edison ran to the neighboring lumberyard and recruited a dozen strong men to carry boards and help him prop up the structure and erect a temporary fence against the wind, which was blowing—in the estimation of one who experienced it—"with the force of a hurricane."

James Craig Watson and Norman Lockyer, meanwhile, made a last-minute decision to gain a few seconds of totality. Rather than observe the eclipse in Rawlins, they would head to Separation, which sat closer to the midline of the eclipse path and therefore would experience a slightly longer phase of darkness. J. B. Silvis, the Union Pacific photographer, offered his wheeled studio for transport. Hooked to the back of a westbound freight train, the caboose carried the two astronomers to the remote rail stop where Edison had hunted the stuffed jackrabbit. Joining them were several volunteers for the day: Watson's wife, Annette; D. H. Talbot, the Sioux City land broker; and the two young men from Cambridge, R. C. Lehmann and his friend James Brooks Close. When the train arrived at Separation, Lockyer erected his equipment by the station, in the lee of the large water tank. Watson, with his wife and telescope, headed on to Simon Newcomb's camp, which sat almost a mile away on the south side of the tracks. Pushing through the thorny brush could not have been pleasant for a man of girth.

IN COLORADO, the people of Denver also awoke to limpid skies. Joseph Brinker, the founder of a private school in the city, kept close track of the weather that morning—at six o'clock, he wrote: "Not a cloud"; seven: "Not a cloud"; seven-thirty: "Not a cloud"; eight: "Not a cloud"—but given the experience of recent weeks, no one could be confident that conditions would remain unchanged in the afternoon.

In the forenoon, locals and visitors prepared for the big event. The eclipse's brief total phase, when the moon would cover the entire surface of the sun, could be viewed safely with the naked eye, but the much longer partial phase required a dark filter for direct observation. To fill this need, Denverites who had been hoodwinked during the recent blue glass craze—sold azure panes to promote their health—now put their poor investment to profitable use; they employed the glass as a solar filter, in some cases fitting it in the bottoms of boxes or the tops of old stovepipe hats. Many children went a different route, collecting shards of clear glass and blackening them over candles. (Neither smoked nor stained glass is deemed safe by modern standards

for viewing the sun, but both were commonly used in the nineteenth century.) "Here's your eclipse glasses," Denver's newsboys yelled, hawking their crude wares for pennies and earning one ambitious youngster a reported seventy dollars over the course of the day.

Some in the Denver area left early for eclipse excursions into the foothills and mountains, taking with them picnics of bread and cheese. Many more scoped out suitable viewing locations in town. Maria Mitchell chose for her observation post, at Alida Avery's suggestion, a hill on the edge of the city, just beyond the reach of suburban development. It was a broad, sloping tract of short grass, easily reached by horse and buggy. Once there, the Vassar party had no time to make elaborate preparations. The women set out wooden chairs, erected a small tent for shade, and mounted their three telescopes on tall tripods. (Mitchell had brought with her the same telescope she had used on her home turf of Nantucket in 1847 to discover her famous comet.) The view east offered an endless, empty expanse of plains. To the west lay Denver, and the Rockies behind it. Immediately to the south sat a three-story brick building topped by a gabled roof and an ornate cross. It was St. Joseph's Home, a Catholic hospital operated by the Sisters of Charity of Leavenworth, Kansas. The nuns in dark habits, spying the astronomers in dresses, came over to offer tea.

The city appeared to be on holiday. As the *Denver Daily Times* had recommended, banks and retail extablishments closed their doors. People gathered on rooftops: the post office, the high school, the fire station, the opera house. A crowd estimated in the thousands assembled along the high ground of Capitol Hill, and in that neighborhood could be found the scientific party sponsored by the Chicago Astronomical Society, including the twenty Denverites who had been specially trained to sketch the corona. They sat themselves on the brow of the hill, facing the sun. A rival team of Chicago astronomers placed itself nearby, on the grounds of the Brinker Collegiate Institute, where principal Joseph Brinker continued to enter notes in his weather log.

Eleven-thirty: "Not a cloud."

Noon: "Single speck of cloud west."

Twelve-thirty: "Three light clouds west."

One o'clock: "Number of small bright clouds west."

A bit over an hour remained until the eclipse began. Looking south from Denver, the growing throngs could see Pikes Peak standing bright and bold against the sapphire sky.

UP ON THE SUMMIT of Pikes Peak, the assembled scientists were at last enjoying sunshine. Samuel P. Langley and his brother spent the morning adjusting their equipment and modifying their observing plans, given that they had lost a member of their team to illness.

That sick participant, Cleveland Abbe, after being evacuated the night before, had been carried not to the base of the mountain but to just below the timberline, where a rustic lodge sat on a lake at an elevation still of about ten thousand feet. At one o'clock in the morning, a doctor arrived to assess Abbe's condition. He ordered Abbe not to return to the summit, and left two nurses to care for the ailing scientist until he was well enough to descend to the base of the mountain. Abbe then scratched out a note to be delivered to his boss at the top of the peak:

My Dear General;

I am most devoutly thankful to you for the good care that you have taken of me—and Dr Hart of Col. Springs whom you have summoned—seems decidedly of the opinion that you have done wisely. I must not oppose my own will to reason & your orders. I will therefore stay here today and organise some sort of system of observing the eclipse so that you shall have a report from the Lake House as well as the summit. . . . I trust that you will not yourself suffer from the Pike Peak "fever"

> I remain yours truly
> Cleveland Abbe.

At daybreak, despite having slept in the somewhat thicker oxygen at slightly lower altitude, Abbe remained weak and faint, yet he was determined to be again what he once was: an astronomer. At noon, he arranged to be carried outside and laid dramatically on a southwest-facing slope with his head propped up. His telescope—a fine instrument made by Alvan Clark & Sons—was still on the summit. All he could rely on were his poor eyes and imperfect spectacles.

ACCORDING TO CALCULATIONS by the Nautical Almanac, the eclipse was set to commence in Rawlins shortly after 2:00 P.M. local time, and in Denver at around 2:20. The event's beginning, like the start of the transit of Mercury, would be barely perceptible—the moon would at first appear like a subtle dent, or flattening, along the sun's western edge. Across the region, everyone watched and waited. The skies held clear, and for those fortunate enough to be in the path of totality, it promised to be quite a show. "[A]t last we were among the favored mortals of earth," one Colorado newspaper remarked.

The rest of the nation was less favored—those outside the shadow path would not witness a total eclipse—but everyone would see at least a partial eclipse, weather permitting. Sidewalk vendors in Chicago, St. Louis, Boston, and elsewhere did a brisk business in eclipse glasses. "Here ye are now," a hawker cried in Manhattan, "blue glass only three cents apiece; all ready to look at th' eclipse—three cents apiece."

In the late afternoon, when the partial eclipse was set to begin in New York, the city's focus shifted upward, as the *Herald* described:

> Portly bankers about to start for home paused on their office steps and turned their eyes above the money making world; merchants stood in the doorways of their busy stores, alternately consulting the face of their watches and the face of the sky; clerks and messengers, hurrying along the crowded streets, ceased to knock and jostle one another and with upturned faces and a blissful forgetfulness of business stood gazing all in one direction, while shop girls, escaping from the toilsome factory, caught a [momentary] glimpse of the heavens above and stalwart policemen stood boldly by frightened French nurses and their infant charges. Even the stage drivers forgot for a single moment to crane their necks and beckon enticingly to passing pedestrians, in the hope of securing another passenger and another fare.

Across the land, as America's attention was drawn to the higher spheres, an otherwise typical workday assumed a new and exotic countenance.

CHAPTER 15

FIRST CONTACT

MONDAY, JULY 29, 1878

2:03:16.4 TO 3:13:34.2 P.M. SEPARATION MEAN TIME

2:19:30.5 TO 3:29:03.5 P.M. DENVER MEAN TIME

"THERE SHE GOES," HENRY DRAPER CALLED OUT THE MOMENT he saw the moon kiss the sun. He was in his pine-shed observatory, in Robert Galbraith's yard beside the Union Pacific in Rawlins. By now a crowd had gathered in the vicinity. The spectators immediately turned toward the southwest, smoked glass in hand, and gazed past the railroad tracks and over the treeless hills toward the blazing sun. It requires a practiced eye to identify the moment when a solar eclipse begins, and the locals probably noticed nothing, but first contact had indeed arrived. Another seventy minutes would have to elapse before second contact, the start of totality.

Across the fence, in the Galbraiths' chicken coop, Edison was still wrestling the whims of a persistent wind. Despite his earlier efforts to shore up the henhouse, his telescope—which projected beyond the confines of the structure—remained subject to the gusts, and they continued to throw his apparatus out of alignment. He now crafted a rigging of ropes and wires to hold the telescope in place. It offered only partial benefit.

The gale was also blowing at Separation, where James Craig

Watson had joined the Newcomb party and erected his telescope in the shelter of a sand ledge that broke most—but not all—of the force of the wind. Soldiers from Fort Steele placed sections of the railway's snow fence on top of the bank to provide added protection, but the fencing tended to blow down.

IN DENVER, WHERE THE ECLIPSE was to start a few minutes later than in Wyoming, the Vassar women spent the moments before first contact watching the sun's edge for the impinging moon. Maria Mitchell, Cora Harrison, and Elizabeth Abbot silently peered through telescopes while Emma Culbertson, holding a timepiece, counted the seconds. Culbertson's nerves got the better of her, and she lost her breath. Professor Mitchell took up the counting until her former student regained composure. When the moon at last made its appearance, the three observers recorded the time. Their estimates varied by more than a second—"a large difference," Mitchell noted.

The beginning of a solar eclipse brings momentary excitement, but what comes next is a lull. Over the following minutes, as the moon nibbles at the sun, one feels the urge constantly to gaze upwards, but with each glance the scene appears virtually unchanged. Given the frustratingly slow pace, one looks for distractions, as did the Vassar party. "Between first contact and totality there was more than an hour, and we had little to do but look at the beautiful scenery," Mitchell remarked.

A photographer was on hand, and the women took time out to pose. They sat in wooden chairs on the grassy plain, bonnets on their heads and hands in their laps, beside the telescopes that aimed up and to the right in precisely the same direction, toward the sun. The image was captured as a stereograph and would be printed on a souvenir card labeled "Colorado Scenery," as if this assemblage of scientists was yet one more marvel of the American West, a wonder to behold like the geysers of Yellowstone or the Rockies themselves.

CLEVELAND ABBE ENJOYED NO team of trained assistants as he lay supine on the slope of Pikes Peak, but he enlisted what help he could as he prepared for totality. The proprietress of the Lake House, one Mrs. Copley, offered him the aid of two young men, who constructed a kind of easel that Abbe could reach without getting up. The men drove stakes into the ground to his left and right. Then they placed on these supports a crossbar onto which they nailed a wooden board that pivoted up and down. It held a sheet of paper. The contraption, which straddled Abbe's recumbent body, would allow him to sketch the corona and to compare his drawing to the real thing in real time, by tilting the board forward and back—toward the sun and away from it. He hoped this would enable him to make an especially accurate and detailed rendering of the sun's mysterious halo.

Four thousand feet higher on the mountain, dozens of tourists had arrived on the Pikes Peak summit. The day-trippers tethered their horses and burros to the rocks, then picnicked beside a snow-bank that they used to chill their wine. The Signal Service had previously cautioned the public, in the press, that the peak was a government reserve and "no one not of the service or connected with its observations, will, without special authority be permit-ted within one hundred yards (100) of the station or any observing party on the day of the eclipse." In reality, the head of the Signal Service, General Myer, proved more liberal than this warning sug-gested. "General, may a woman come within your charmed circle on the peak?" asked Mary Rose Smith, a visitor from Philadelphia. "I invite you, madam, on condition of perfect silence when nearing the totality," he replied.

The stipulation agreed to, Mrs. Smith and her small party took a seat on the northern edge of the summit, at General Myer's sug-gestion, and faced in the direction of Rawlins. That part of Wyo-ming was far off in the distance, hidden behind several ranges of snow-flanked peaks, but the vista would be ideal for seeing something else: the approach of the moon's shadow in the moments before totality. In the meantime, the tourists watched the eclipse progress through colored glass. As the moon carved an ever-larger bite out of the solar disk, the sun grew moonlike: a shrinking crescent.

A marmot—a large rodent with a bushy tail and an inquisitive face—emerged from the rocks below. It squatted and grinned.

FOR MUCH OF THE SPAN of the partial eclipse, spectators may well have wondered if the hype had been justified. The sun was one-quarter covered, then half covered, then three-quarters covered, and still there was no noticeable reduction in daylight. If astronomers had not alerted the public to what was happening overhead, many people would have gone about their lives unaware. But about fifteen minutes before the moon completely blocked the sun, strange things began to happen.

In the parlor of Denver's American House hotel, a woman marveled at a narrow sunbeam that shone on the floor, painting the shape of a sickle. It was an image of the eclipsed sun projected through a puncture in a window shade, the tiny gap serving as an inadvertent pinhole camera. The same phenomenon could be seen in nature. Beneath trees, where the sunlight filtered through a thick layer of leaves, what dappled the ground were not points of light, but crescents. The display under the cottonwoods

CRESCENTS VISIBLE UNDER FOLIAGE
DURING PARTIAL ECLIPSE

that lined the streets of Boulder presented, as one local put it, "a gorgeous appearance which would have delighted the heart of a Turk."

Another effect, less obvious to identify, gave the landscape an otherworldly appearance. On a normal cloudless day, the shadows of people and buildings wear ragged edges. This blurry outline, the penumbra—a zone between full shade and full light—results from the sun's considerable width in the sky, which causes its rays to arrive on earth at a range of angles. Now, as the moon covered the sun, the solar surface effectively shrank. The sun's rays became more parallel, shadows grew sharper, and objects seemed more clearly defined, as if displayed at high contrast. As the solar radiation decreased, so did the temperature. In Denver, one visiting astronomer, already wearing a light coat on what had been a warm summer's day, called for a second jacket. To the west, on a high mountain pass, men and women who had ascended for the spectacle wrapped themselves in cloaks and blankets.

Meanwhile, the sky's hue began to shift. Francis Cranmer Penrose, an English architect and astronomer who had observed the 1870 eclipse in Spain, was struck by the changing light around Longs Peak, a sheer promontory to the northwest of his observing post near Denver. He watched the mountain and the sky behind it turn lilac, while the upper sky deepened to "a very dark warm blue." In Rawlins, one observer noted a "strange, weird, grey light, resembling that which precedes the dawn." Mary Rose Smith, the tourist from Philadelphia, perceived a similar change in the landscape's complexion as she gazed north from the summit of Pikes Peak. "The light of the sun grew pale and grey," she wrote. "All the yellow rays seemed to fade out of it, and the face of nature and of man took on a weird and ghastly palor [*sic*]."

About eight minutes before totality, a volunteer working with British astronomer A. C. Ranyard at the Princeton camp pointed to an object in the sky. It was Venus, "which was shining brightly, and might no doubt have been seen some time previously if we had been looking for it," wrote Ranyard. "The heavens had assumed a violet tint which each minute was growing deeper."

Animals perceived the ebbing of the light, and they responded as they normally would at the close of day. In Rawlins, owls emerged. Farther north, in a Montana gold mining town, "all the cocks in the city began to crow lustily and in a regular succession." Across the region, cows turned homeward and pigeons went to roost. Grasshoppers folded their wings and fell to the ground.

WITH TOTALITY RAPIDLY APPROACHING, tension escalated in the scientific camps. Astronomers demanded that spectators remain silent, and not just on Pikes Peak. In Denver, the mayor stationed a police officer at the Chicago Astronomical Society's observation post to maintain public order. In Rawlins, as one newspaper reported with presumed exaggeration, the Draper party ordered "that if, during the totality, anybody came near or spoke to them, 'shoot him on the spot.'"

Inside the Draper observatory, all was accordingly quiet. Henry Morton was not there; he had left to make his observations from atop a nearby hill. The three who remained—Henry and Anna Draper, and George Barker—prepared candles in case they needed the light to take notes after the loss of the sun. They appeared ready for totality. "The only place of disorder was in that frail structure of Edison's," wrote *The New York Herald*'s Edwin Marshall Fox. "Notwithstanding his efforts the wind continued to give him trouble. In vain he adjusted and readjusted [the tasimeter]. . . . Edison's difficulty seemed to increase as the precious moment of total eclipse drew near."

Up the tracks at Separation, James Craig Watson prepared to make his quick search for Vulcan. Even before the sun disappeared, he began sweeping the heavens for any bright objects that had become visible in the darkening sky. Simon Newcomb, meanwhile, was hiding in the camp's photographic darkroom to sensitize his vision. (Other astronomers, for the same effect, bandaged their eyes.) Newcomb emerged just three minutes before totality. By now, the sun was a mere sliver. As he made his way to his telescope, he noted the "lurid" color of the landscape. "The light seemed no longer to be that of the sun, but rather to partake of the character of an artificial illumination."

With just a minute to go before totality, another bizarre phenomenon became visible to some. As if the sun were being projected through shallow water at the beach, narrow bands of light and shade rippled across the ground, or—from the viewpoint of astronomer Edward Holden, who was stationed atop the Teller House Hotel in Central City, Colorado—across the roof. "They coursed after each

SHADOW BANDS OF 1870 ON
AN ITALIAN DWELLING

other very rapidly," he wrote, "seeming about 3 feet from center to center, the dark band being, say, 6 inches wide, the interval being bright." These wavy lines, termed *shadow bands*, are not always seen but can be dramatic, as at the total eclipse of 1842 in Southern France, where the undulation was reported to be so striking that "children ran after it and tried to catch it with their hands." The cause of these ripples is the same that makes stars twinkle—currents of warm and cold air that bend light as it passes through the atmosphere. Indeed, shadow bands have been called, poetically, "visible wind."

The sun's crescent had now grown exceedingly slender, a mere filament. It continued to shrink, like an ember burning itself out at the ends. Before vanishing, however, this glowing thread produced a final brilliant display. It shattered into a string of shimmering jewels. These dancing points of light, called *Baily's beads* (described and explained by British astronomer Francis Baily in 1836), are the last of the sun's rays filtering through valleys on the edge of the moon.

In the closing seconds before the onset of a total solar eclipse, darkness falls with disorienting rapidity. It can feel as if you are losing your eyesight, or perhaps your sanity. The dimming light does not just surround you; it swallows you. The very ground seems to give way.

In the midafternoon on July 29, 1878, as the people of southern Wyoming plunged into shadow, they withdrew the smoked glass from their eyes and beheld a sky like none they had seen before.

CHAPTER 16

TOTALITY

MONDAY, JULY 29, 1878

3:13:34.2 TO 3:16:24.2 P.M. SEPARATION MEAN TIME

3:29:03.5 TO 3:31:44.0 P.M. DENVER MEAN TIME

A TOTAL ECLIPSE IS A PRIMAL, TRANSCENDENT EXPERIENCE. The shutting off of the sun does not bring utter darkness; it is more like falling through a trapdoor into a dimly lit, unrecognizable reality. The sky is not the sky of the earth—neither the star-filled dome of night nor the immersive blue of daylight, but an ashen ceiling of slate. A few bright stars and planets shine familiarly, like memories from a distant childhood, but the most prominent object is thoroughly foreign. You may know, intellectually, that it is both the sun and moon, yet it looks like neither. It is an ebony pupil surrounded by a pearly iris. It is the eye of the cosmos.

The sight, for many, is humbling and mystical. A Princeton student who witnessed totality in Iowa in 1869 compared the rush of emotion to an earlier near encounter with death, when "I was once held in a drowning condition at the bottom of a stream and the review of my life passed before me." The hypnotic effect appears to extend beyond humans. On the eve of the eclipse of 1842, in Southern France, a local man—for an experiment—withheld food from his dog. "The next morning, at the instant when the total eclipse was going to take place, he threw a piece of bread to the poor animal, which had begun to devour it, when the sun's last rays disappeared," the astronomer François Arago recounted. "Instantly the dog let the bread fall; nor did he take it up again for two minutes, that is, until the total obscuration had ceased; and then he ate it with great avidity."

Scientists, too, are apt to be spellbound. Firm hands tremble, eloquent tongues freeze, sharp minds grow addled. "In fact, the general scene of a total eclipse, is a potent Siren's song, which no human mind can withstand," warned Piazzi Smyth, the Scottish Astronomer Royal. "For its effects on the minds of men are so overpowering, that if they have never had the opportunity of seeing it before, they forget their appointed tasks of observation."

Now, in Wyoming, the "darkness was like that of deep twilight, and the beauty of the corona was enchanting, but there was no time to pay attention to such things," wrote William Harkness, head of the government party at Creston. "[E]verybody worked as if for their lives."

TOTALITY WAS PREDICTED TO last 176 seconds at Creston. At Separation, the duration would be five seconds shorter, and at Rawlins another nine seconds shorter still. Every moment in the moon's shadow was precious. Keeping track of time was essential.

The scientists—occupied at their telescopes, spectroscopes, and cameras—had many tasks to complete before the sun's reemergence; they could not spare the distraction of watching a timepiece, so each camp assigned an assistant to count the seconds aloud. For the Draper party in Rawlins, that job fell to Anna Draper, who had filled a similar role for her husband during his observations of the transit of Mercury. Norman Lockyer, beside the Union Pacific station at Separation, gave the duty to R. C. Lehmann, the young man from Cambridge. A short distance away, at Newcomb's settlement, the timekeeper was a soldier from Fort Steele, Captain William H. Bisbee. The wind howled across the arid plain, drowning out voices, so Bisbee hammered against the round cover from a camp stove. On

TOTAL SOLAR ECLIPSE, JULY 29ᵀᴴ 1878.
Simon Newcomb; Separation, Wyoming.

every tenth beat, he called the time.

While the seconds rang out, as if from a resonant metronome, Newcomb began his work by examining the faint outer corona with the naked eye. To view these delicate tendrils clearly, he hid the brighter, inner corona behind a circular screen he had set atop a telegraph pole. (In essence, he eclipsed the eclipsed sun.) He was astonished to see how far the corona extended, stretching many times the sun's diameter to the upper left and lower right. Nearby, an assistant from the Naval Observatory used a spectroscope to analyze the corona, while a colleague observed through a telescope. Several lay volunteers were also there: James Craig Watson's wife, Annette; D. H. Talbot, the Sioux City land agent; and Wyoming's newly appointed territorial governor, John W. Hoyt, a former chemistry professor and a staunch advocate of scientific research, who, like Watson, had served as a judge at the Centennial Exhibition in Philadelphia. These three acted as artists, sketching the corona in pencil on sheets of white cardboard.

Watson, meanwhile, stood at the edge of the group, beside the sand ridge that offered shelter from the gusts. He set to work quickly and systematically, necessarily so if he had any hope of finding

Vulcan in less than three minutes. Spotting a world close to the sun might seem a simple proposition, but skilled astronomers had looked during previous eclipses, without luck. (Simon Newcomb confessed to his diary after the 1869 eclipse in Iowa, "Was disappointed in not finding intra-mercurial planets.") Presumably Vulcan was not very bright; it must blend in as just another star. Watson kept a star chart on hand for reference, in case he spotted something that seemed out of place, but he hardly needed it. He had memorized the pattern of known stars he should see near the eclipsed sun.

Captain Bisbee announced the elapsed time. *Ten seconds.*

Watson pointed his telescope at the sun and then gently turned his instrument toward the east. He used a low-powered eyepiece that enabled him to scan a large area in a short time. His field of view moved through Cancer. He recognized several stars, including one that fell in the middle of the constellation and was called Delta Cancri by its Bayer designation. (German astronomer Johann Bayer's celestial atlas of 1603 introduced a simple nomenclature of the heavens, labeling the most prominent stars in each constellation with Greek letters, usually in order of brightness. Delta Cancri means simply the "D" star in Cancer.) Watson saw nothing unexpected. He reset his telescope, back toward the sun, then adjusted it downward to take in a new strip of the sky. He scanned east again, and again he saw nothing unusual.

The metallic beats on the camp stove continued to resound. *Twenty seconds.*

Watson pointed the telescope at the sun once more, then began scanning to the west.

EDISON, MEANWHILE, KEPT STRUGGLING with his tasimeter. After months spent publicizing the invention and weeks perfecting it, he now had less than three minutes to prove its value to science— and *The New York Herald* was watching. Finally, at the moment he most needed it, Edison received a lucky break.

A total eclipse obstructs not only the sun's direct light, but its heat, which can cause an abrupt change in atmospheric conditions— sometimes turning a calm day unsettled; at other times, the opposite. The eclipse of July 28, 1851, for instance, fell on a blustery afternoon in Norway. "The most singular thing," one observer noted, "was, that the wind, which . . . blew hard in the early part of the eclipse, totally ceased, leaving the lake below me smooth and calm as glass." This was now the fortunate effect in Rawlins. As Edwin Marshall Fox of the *Herald* wrote, "Totality had brought with it a marked cessation in the force of the wind."

The sudden stillness was good news for the entire Draper party. In the main observatory, Anna Draper counted the seconds while her husband split the coronal light into its spectrum and photographed the result with a contrivance he called a *phototelespectroscope*. George Barker took a few seconds to examine the corona with the naked eye, looking for prominences and gauging its brightness. "The amount of light seemed to be nearly or quite equal to that given by the moon when ten days old," he estimated. He then turned to his own spectroscope.

The instruments of Draper and Barker "were in excellent working order," *The New York Herald*'s Fox reported. "Still Edison's tasimeter was out of adjustment." The contraption, sitting in its double tin case of water for insulation and connected by wires to a battery and a meter for measuring electric current, had to be aligned precisely with the telescope. Edison needed to project light from the corona through the tasimeter's tiny slit and onto the hard rubber inside. In essence, he had to take a beam of energy that stretched 93 million miles from the sun to the earth and thread it through a gap finer than the breadth of a hair, and he had to do so while merciless time continued its swift progression.

ONE MINUTE THIRTY SECONDS.

Back at Separation, totality was more than half over, and Watson was making his fifth sweep with the telescope when, near the star Theta Cancri, his eye hit something unusual. It was a reddish object, located where no star should be. Perhaps it was a comet, he thought. He looked for elongation, because a comet that close to the sun should have a tail, yet he saw none. His mysterious object did not look like a star, either, for it did not appear as a mere point of light but rather displayed—as Watson perceived it—a noticeable diameter. It was a planet, he was sure. His heart must have leapt. He had found Vulcan.

One minute forty.

Watson, however, could indulge no momentary celebration. His discovery would mean nothing if he could not specify where in the sky the planet resided. In the dim light of the eclipsed sun, he took up his pencil. Using the contraption he had attached to the telescope, he marked where the brass pointers fell on the wooden circles. He would later read these marks to specify Vulcan's coordinates.

One minute fifty.

AT ALMOST THE SAME MOMENT that Watson nabbed his elusive planet, Edison succeeded in aiming the coronal light through the slit in his device. He watched intently.

The electrical setup of his tasimeter had not been working quite right. In its null condition—with no heat applied—the needle should have sat still. Instead, it had been moving slowly leftward, toward "cold." As soon as the coronal light landed on the rubber strip inside the device, however, the needle stopped moving. Then it started moving right—toward "heat." It accelerated. Within five seconds, it shot off the scale entirely.

Edison was "in ecstasies," according to one who saw him a short time later. The tasimeter had found heat in the corona—indeed, more than Edison had expected. Had he anticipated so much heat, he would have adjusted his device to be less sensitive so as to get a precise reading. With less than a minute of totality left, however, there was no time to rerun his experiment. Besides, the wind picked up again. The henhouse began to totter.

DURING THE FINAL MINUTE of the total eclipse at Separation, Watson continued his search for planets west and south of the sun. With just about twenty seconds to spare, he stumbled on another object he could not definitively identify. He assumed it was a star— actually, a tight cluster of stars—called Zeta Cancri, but with totality almost over, he had no time to check its placement against other stars in the vicinity. Instead, he quickly marked its location with the brass pointers on the wooden circles. He would later compare its position to known stars on his chart.

Two minutes forty seconds.

With the sun about to reappear, there was one thing Watson still hoped to do: gain independent confirmation of his discovery of Vulcan. He ran a few yards to Simon Newcomb, who by now was at his own telescope sweeping for unknown planets, and urged him to turn his glass toward the ruddy point of light southwest of the sun. Newcomb, however, could not comply; he had found his own strange object and was busy reading its position in the sky. (Newcomb would later determine that his object was nothing more than a known star, and he would lament, "It is of course now a matter of great regret that I did not let my own object go and point on Professor WATSON'S.")

Then, with jarring suddenness, the show came to a close as a blinding point of sun emerged from the trailing edge of the moon. Daylight returned, seemingly faster than it had gone. The stars and known planets, the corona, and Watson's mysterious object speedily vanished behind the brightening scrim that is the earth's blue sky.

EVEN AS TOTALITY ENDED in Wyoming, it had not yet begun farther south, in Colorado. The moon's shadow raced toward Denver at a bullet's clip, some two thousand miles an hour. "Those who have seen a locomotive approach . . . can judge of the stupefaction caused by the approach of this black column with all but lightning speed," Scottish glaciologist James D. Forbes wrote after watching the lunar shadow dash across the Alps in 1842. "I confess it was the most terrifying sight I ever saw." Now, in the Rockies, the tourists who had gathered on the northern lip of the Pikes Peak summit saw what appeared to be a monstrous, silent storm rushing forward. It was "a solid, palpable body of darkness, rising up in a great wall," as Mary Rose Smith described it. A companion called it "an angry black cloud of inky blackness . . . advancing with startling rapidity as though bent on destroying all before it." The impenetrable curtain engulfed the distant peaks. One by one, those lit by the sun's last rays popped out of sight.

As darkness fell, "cheer after cheer echoed and re-echoed among the surrounding mountains," observed *The Denver Daily Tribune*. Even those spectators who had pledged to remain silent so as not to disturb the astronomers could not help themselves. "[W]hen the total obscuration was reached, one great universal shout which took in all the great heart of Denver, went up from the streets and house-tops," *The Chicago Times* reported. The public marveled at the corona and the weird, dim light it cast. "It permeated everything; the faces of the observers partook of the ghastly glamor; the habitations of stone, brick, and wood were all tinted with a hue that seemed to come from another world."

On Denver's Capitol Hill, the corona-drawing class organized by the Chicago Astronomical Society was laboring with intensity and purpose. Each volunteer drew in pencil his or her pre-assigned portion of the strange halo around the sun. Together, the sketches revealed an asymmetrical, plumed appendage—a mismatched pair of swallowtails shooting out on opposite sides of the sun. But even this composite view could not be considered a definitive picture. Perceptions of the corona are remarkably subjective; two people standing beside each other will often see a different shape and texture, and members of the class could not agree on the corona's color. Three of the volunteer artists described it as yellow. Two called it white. One said orange. Four others answered that it was the color of straw.

The sky, too, was of a perplexing and ever-changing hue. F. C. Penrose, the British scientist who had noted a lilac backdrop to Longs Peak in the minutes before totality, now marveled at the bright yellow-orange horizon beyond the mountain—the impending dawn that would follow the afternoon's brief night.

At the Princeton camp along Cherry Creek, not all was going according to plan. Charles Young had long since abandoned the tasimeter he borrowed from Edison—he simply could not get it to work right—and the Englishman A. C. Ranyard, who had been among those shipwrecked on the way to Sicily in 1870, was about to hit some metaphorical rocks. He had traveled five thousand miles to Denver with a bulky, large-format camera intending to photograph the corona in exquisite detail. The instrument required several men to operate, so Ranyard had recruited three local assistants who spent five days conducting practice drills, until their duties had been rehearsed to seeming perfection. On the day of the eclipse, one member of the team—he happened to be Denver's superintendent of public schools—invited his thirteen-year-old son to help. During totality, Ranyard thought all was functioning well with the camera, but shortly

Camera in position

after the eclipse he would discover—to his dismay—that most of the exposures captured no image. He would blame the mishap on the lad, who when handing unexposed glass slides to his father for insertion into the camera had apparently moved some critical hooks. "If I have the good fortune to observe another total eclipse, and should be compelled at the last moment to accept the assistance of an intelligent boy," Ranyard would later relate in this cautionary tale, "I shall be careful to explain to him the object of any hooks or clamps that may come in contact with his fingers before totality commences."

ON DENVER'S EASTERN EDGE, the Vassar party did not attempt anything technically complex during totality. The three women at telescopes—Maria Mitchell, Cora Harrison, and Elizabeth Abbot—examined the corona's shape and color, and searched for unknown planets. The others made naked-eye observations of the landscape and sky. The women saw Mercury, Mars, and Venus. They found no Vulcan.

For this group of observers, however, viewing the eclipse was arguably less important than *being* viewed. The Vassar women, far from the sexless Amazons that Dr. Clarke had warned would result from female higher education, presented irrefutable, concrete evidence that science and femininity could coexist. These astronomers in pleated dresses provided "an attraction to the gaping, yet respectfully distant, multitude of masculines, almost as absorbing as the eclipse," a reporter wrote. "PROF. MITCHELL HERSELF, as with iron-gray curls fluttering under a broad-brimmed Leghorn, she swept the heavens with a four-inch telescope, or directed with native majesty and grace the operations of her assistant nymphs, was a figure, and perfectly commanding." The Vassar astronomers also proved inspirational to members of their own sex. "[W]omen of low and high degree throughout the territory turned during that day their thoughts toward the hill, even as the pilgrims of old prayed with their faces toward Jerusalem," wrote another correspondent, "for from the mound where the group stood there radiated a light, that sent its rays hopefully into more than one woman's heart—a heart with longings for study, culture, improvement, that the simple fact of her being a girl had unjustly deprived her of because old prejudices had hedged her path and defined her duties."

During the eclipse, the Vassar expedition served as a kind of political theater, promoting social change, and it could hardly have found a more consequential audience. Across the way, near the hospital run by the Sisters of Charity, three Jesuit scientists had established an observation post. Just as the eclipse began, they were joined by Denver's bishop, Joseph P. Machebeuf, the outspoken opponent of women's suffrage who only a year earlier had preached, "Women are not needed as men; they are needed as women, not to do what men can do, but to do what men cannot. Woman was created to be a wife and a

mother." What Machebeuf made of the Vassar astronomers, who in the dim, unearthly light of the corona must have looked like beings from another world, is not known. His reactions have been lost to the ages.

FOR CLEVELAND ABBE, flat on his back on the shoulder of Pikes Peak, the 161 seconds of totality felt surprisingly leisurely. All he planned to do—indeed, all he could do in his condition—was examine the shape of the corona and trace its contours. At his only previous total eclipse, in Dakota Territory in 1869, he had used a telescope, which allowed him to examine the corona in detail in the immediate vicinity of the sun. His companions who had used no magnification, and whose eyes captured a much larger portion of the sky, told him they had observed immense coronal streamers that stretched millions of miles into space. Abbe always assumed that what his colleagues had seen were optical illusions, perhaps caused by the way the light was transmitted by the earth's atmosphere, but now he saw the same thing—long, glowing shafts that projected from the sun in various directions. He tried to determine if they were illusory. He turned his head to see if the rays shifted their position. They did not. He looked for any subtle motion or flicker. He saw none. As far as he could tell, what he was looking at were actual, enormous structures in space.

While struggling to focus his eyes through his spectacles, Abbe examined the region around the sun over and over again. As his vision adapted to the darkness, he discerned ever more detail. He estimated the length and width of the rays, and the angles between them, and he drew what he saw. It became increasingly clear to him that this seeming multiplicity of rays was really just two light beams that crossed the sun nearly at right angles,

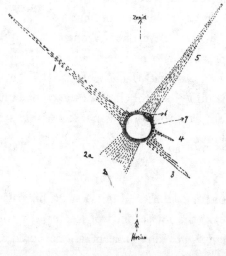

each beam wide at one end and pointed at the other. "I became perfectly convinced that these rays were permanent and real, and more important than I had hitherto believed," he concluded. He was still faint from the altitude, and his head ached from straining his eyes and taxing his brain. He puzzled over what he was looking at.

———

UP ON THE SUMMIT of Pikes Peak, General Myer, a man usually disdainful of scientific research, seemed now to fancy himself an astronomer. Taking charge of Cleveland Abbe's telescope (much as he had previously commandeered Abbe's nickname), Myer pressed his eye to the glass and noted what he saw: "No red color in corona; absence of prominences." Meanwhile John W. Langley, the chemistry professor, used an improvised *photometer* to measure the brightness of the corona. His brother, however, found himself with remarkably little to do. Samuel P. Langley's original plans had been to use a tasimeter and a spectroscope, but Edison had failed to send the former, and the latter had to be left at Allegheny when the U.S. Naval Observatory delayed in shipping a telescope to which it was to be attached. And so, the one true astronomer atop Pikes Peak spent most of totality doing what a team of amateurs was doing in Denver, sketching the corona while viewing it with the naked eye. He was not disappointed, however, to perform so little science. Langley had previously confided to Cleveland Abbe a secret wish—"to *see* the eclipse (I have 'observed' two but not seen any as a spectator)"—and a week after the event he would write of its visceral impact: "I once experienced an earthquake, and I think this and a total eclipse of the sun are two things that it is no use trying to describe; you must feel or see for yourself."

The tourists who had ascended Pikes Peak—those who had witnessed the approach of the moon's shadow a few minutes earlier—now peered north for the return of daylight. Gazing toward Wyoming, they watched as the distant mountaintops that had been shrouded in darkness reappeared one by one until, suddenly, it was they who were back in the sunlight. Samuel Langley cried "over." Totality had ended.

Across the region, cheers once again rose from the mountains, the foothills, the streets, the rooftops. On Grays Peak, a high summit west of Denver, two shots rang out as a member of a large party fired his revolver in celebration. "Then every tongue was unloosed. The ladies started 'My Country 'tis of thee,' and sang it with a will," reported one among the crowd.

The Vassar women, in contrast, were silent—privately recording their impressions of totality—when someone interrupted with a shout: "The shadow! The shadow!" Maria Mitchell turned her head and caught the great shaft of gloom as it swiftly departed. To a sheepherder south of Denver, the shadow "looked like a black carpet sliding over the plains." To those viewing from above, on Pikes Peak, it was "a rounded ball of darkness with an orange-yellow border fading into the light pea-green of the landscape." They watched as it rushed toward the remote horizon, in the direction of Texas, then seemed to lift off the earth and recede into space.

CHAPTER 17

AMERICAN GENIUS

JULY 30—AUGUST 27, 1878

I T WOULD TAKE MORE THAN A WEEK FOR PAPERS OUTSIDE OF Texas to relay the disturbing news of what had occurred when the moon's shadow reached the Lone Star State. "[T]here were thousands of ignorant people, both white and black, who had not heard that anything peculiar was about to happen," reported the *St. Louis Globe-Democrat*, which included details of the father-son murder-suicide in a salacious roundup of vice and crime headlined SATAN'S CARNIVAL. "A terrible tragedy in Johnson County may be set down to the eclipse," the article continued before recounting the dreadful acts of one Ephraim Miller. "When the eclipse commenced and the darkness of totality came on he ran from the field to his house with a hatchet in his hand," the paper noted. "[W]ishing to take his ten-year-old boy with him to the other side of Jordan, [Miller] raised his hatchet and split his son's head open. Leaving the latter weltering in his blood and struggling in the last throes of death, the father, on a ladder, ascended to the top of the house. Here, with a new razor, he cut his throat from ear to ear, and fell to the ground a corpse."

News outlets in Chicago, Boston, Salt Lake City, and elsewhere soon printed horrifying accounts of the tragedy, irresistible to an American press that often catered to readers' base interests; but the lurid tale, tucked in the back pages, was a mere postscript. The main story of the American eclipse of 1878, as it was written in the event's immediate aftermath, proved an uplifting narrative, a tale of a nation's enlightenment and undeniable progress.

"PROFESSOR WATSON FINDS VULCAN," a Wyoming headline proclaimed the day after the eclipse. Word quickly spread by wire and post. "A new planet discovered, and the discoverer an American!" wrote D. H. Talbot, the Sioux City land agent, in a dispatch back home. "Englishmen may come among us, and endeavor to search out the wonders of the firmament, but they cannot get ahead of American genius." The most influential of those visiting Englishmen, Norman Lockyer, editor of *Nature*, recognized the American achievement in a letter to London's *Daily News*. "[L]ittle doubt remains that a new major planet has been discovered," he wrote. "Professor Watson's work has been acknowledged on all hands to be a veritable *tour de force*."

Some astronomers, largely in private conversation, questioned Watson's discovery—after all, others who had searched for Vulcan during the eclipse failed to find it—yet the prevailing opinion was summed up by the *Boston Daily Advertiser*, which called Watson "a thoroughly competent and trustworthy observer. Indeed," the paper continued, "it may be said with confidence that his keen eyes were not likely to be deceived, or that he could have fancied he saw a planet which was not in view."

The subsequent hoopla draped Watson in adulation. A man not short on ego before the eclipse, he surely accepted as fact what one newspaper now called him—"the most noted astronomical observer and discoverer in the world." Another paper soon made an inevitable comparison when it wrote of Watson, "He is the Edison of astronomy." On his return journey to Ann Arbor, while passing through Chicago, Watson—like Edison a few weeks earlier—found himself cornered by a reporter.

"What were the circumstances of your discovery of Vulcan?" the journalist wanted to know. Watson offered a technical response. "I found a star of four and a half magnitude [in brightness], which had a decided disk, and was in a position in which there is no known star," he said.

The reporter clearly hoped for a more engaging quote, and he devised a way to break through the professor's decorum. The hunt for asteroids was often likened to a celestial round of billiards, and in the American rules of the game, one player might grant the other a handicap—called a *discount*—whereby an opponent's pocketing a ball forced the player to deduct points from his own score. The journalist joked to Watson, "You probably will want some discount for this in the race you and Prof. Peters are having, will you not?"

Watson laughed. "Well, I suppose so. I wouldn't take a dozen asteroids for this."

FOR MARIA MITCHELL, the eclipse had produced no great scientific discoveries, but her expedition too had achieved a remarkable goal. "The success of this party is one more and pointed arrow in the quiver of woman suffrage argument and logic," wrote a correspondent for the New York *Sun*. The Denver press gauged her accomplishment even more generously. "Recently, here in our midst, a conspicuous example of the power and grasp of the feminine intellect has been exhibited," effused the *Rocky Mountain News*. "We allude to Miss Mitchell, and the great interest she is exciting as a scientist. . . . In this she has done a service which all the women's rights pleaders on the continent could never dream of accomplishing."

Mitchell wisely harnessed the attention to further her cause. On the day after the eclipse, she offered a public lecture at Denver's Lawrence Street Methodist Church, the proceeds to support the Colorado Equal Rights League, a new organization to promote equality of the sexes after the recent state suffrage defeat. Mitchell, "the most distinguished lady astronomer of the age," as one Denver newspaper called her, told her audience of a distinguished lady astronomer from a

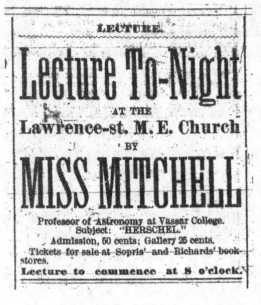

previous age: Caroline Herschel. The younger sister of the German-born British astronomer Sir William Herschel, who was best known for his discovery of the planet Uranus in 1781, Caroline had long been her brother's helper—sitting up all night to record his observations of double stars and cloud-like nebulae, and spending days compiling the data, making calculations, and organizing it all into a vast catalogue of the heavens. Caroline's painstaking work earned her honors from scientific societies and a salary from King George III (although, Mitchell was wont to note, "when [the king] found that she was doing a man's labor he gave her a woman's half-pay"). Mitchell contended that Caroline's innate talents "were as peculiar and as plainly marked as were those of her brother," and yet, as a female assistant, she subordinated her needs and never reached her full potential. "If what she did is an example, what she did not do is a warning," Mitchell admonished. The frontier audience seemed to appreciate the lesson. At the end of the speech, admirers presented Mitchell with an elaborate flower arrangement, shaped like an M and bordered in the colors of Vassar College—rose and gray, symbolizing sunrise through the pall of women's long-suppressed intellectual lives.

THOMAS EDISON, MEANWHILE, a man who hardly needed additional press attention, reaped the harvest of his adventures in the

West and in the erudite realm of pure science. On the day after the eclipse, *The New York Herald* ran Edwin Marshall Fox's account of the Draper party's activities. "The observation of the eclipse has been a grand success and the astronomers here are in a high state of happiness," Fox had telegraphed from Rawlins, offering dramatic details of the windstorm that nearly derailed the experiments. "Edison's observatory, which, in its normal condition, is a hen house, was particularly susceptible," he wrote. "Every vibration threw the tasimeter into a new condition of adjustment." Fox told a gripping tale of Edison's wresting victory from the forces of nature. "At last," Fox wrote, "just as the chronometer indicated that but one minute remained of total eclipse, he succeeded in concentrating the light from the corona upon the small opening of the instrument. Instantly the [needle] on his graduating scale swept along to the right, clearing its boundaries. Edison was overjoyed."

As with all stories about Edison, this one spread nationally, if not globally, although many journalists were at first unsure how to interpret the news. *The New York Times*, for instance, noted both that "Edison's experiments with the tasimeter were quite satisfactory" and, two paragraphs later, "Edison's tasimeter failed to work satisfactorily." The Washington, D.C., *Evening Star* sought to clear up the confusion. "Edison's trip to the west has resulted in the discovery that there is considerable heat in the sun's corona," it explained. "His tasimeter was, however, too delicately adjusted, and the unexpected heat could not be accurately measured. So the experiment was both successful and unsuccessful as reported, with seeming contradiction, in the dispatches."

Although Edison could not say exactly how much heat came from the corona because his instrument's needle had flown off the scale, this flaw in the experiment added ironic luster to his invention. The tasimeter had proved acutely delicate, so much so that Edison suggested a new application of his device—to scan the night sky for invisible celestial objects that are detectable only by the minuscule heat they emit. "His plan is to adjust his tasimeter to its extreme degree of sensitiveness and attach it to a large telescope which moves slowly in a semi-circular direction," Fox explained in a follow-up story in the *Herald*. "In this way he states it will be possible to discover stars which are too remote to be seen. In other words, when he cannot see them he will feel them." *Scientific American* praised the concept, and although the tasimeter never would be used in this manner, the idea was indeed ahead of its time. Edison had, in fact, anticipated the development of infrared telescopes, devices that a century later would allow astronomers to peer through interstellar dust clouds and to uncover hidden galaxies in the deep recesses of the universe.

Edison did not return directly to Menlo Park from Wyoming. He and George Barker took a long, meandering route home, first journeying west on an extended vacation to San Francisco, Yosemite, and Virginia City, Nevada, then stopping back in Rawlins for their prearranged fishing trip with Major Thornburgh of Fort Fred Steele. Finally, heading east, the pair stopped in St. Louis for the annual meeting of the American Association for the Advancement of Science. The gathering of academics and scientifically minded laypeople included several individuals who had also observed the eclipse and some of the same scholars Edison had encountered in Washington at the National Academy of Sciences. Simon Newcomb, the association's outgoing president, opened the assembly—to meager attendance due to a yellow fever epidemic in Louisiana and Mississippi that, given its potential spread to Missouri, had scared visitors away. "The meeting here, will I fear, be the smallest and dullest ever held," Newcomb brooded in a letter to his wife.

Edison's arrival, however, infused the conference with much-needed pizzazz. "Edison is undoubtedly the 'big gun' of the meeting," wrote one in the audience. "It was curious to observe how even the men whose eyes had grown dim, and whose hair had grown gray, in scientific investigations, listened with almost reverential attention to his technical explanation of 'The use of the Tasimeter for measuring the heat of stars and of the sun's corona.'" The lecture, presented just after noon on the third day of the conference, was an awkward address. Given Edison's deafness, nervousness before large groups, and lack of preparation ("I haven't written the paper yet," he admitted the previous day), the famed inventor spoke quietly and haltingly while gesturing toward a chalkboard and disassembling a tasimeter to show its inner workings, then concluded abruptly, "I believe that is all I have got to say." The packed audience at Washington University's chapel was awed nevertheless, applauding heartily and asking many questions. Earlier in the meeting, attendees had formally elected Edison a member of their scientific body, and that morning, the association's incoming president, Yale dinosaur specialist O. C. Marsh, had publicly praised Edison for his work that "reflected glory upon the progressive genius of America." The unschooled inventor bowed and blushed. A headline deftly summed up the day's events: THE GREAT EDISON. THE SCIENTISTS WELCOME HIM TO THEIR RANKS.

The following Monday, August 26, Edison arrived in Menlo Park, tanned and rested after a month and a half away. As he sat in his workshop, his young daughter danced up to him and tapped his shoulder. "Why, Dot, is that you?" Edison cried, showering her with kisses. Soon, though, he shifted his attention from his family back to the press. Several journalists had come to report on the Wizard's return, and Edison regaled them with lively anecdotes from the journey—the stuffed jackrabbit, the tottering henhouse, the trout so abundant you could catch them without bait. He even demonstrated his newfound marksmanship, taking out his Winchester and shattering a glass insulator placed on a post at a distance.

"Did you get any new ideas out there, Mr. Edison?" a reporter asked, inquiring about the lasting impact of the trip to the West.

"No," Edison said. "That's not a place for ideas. It's perfectly barren; but it's a splendid country. I'd like to go out there every summer."

The next day, in his laboratory, Edison began work on a new project, one that would propel his fame, wondrous as it already was, to unimaginable heights.

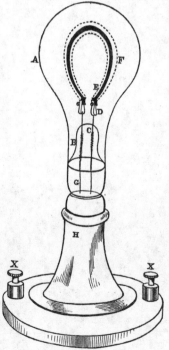

CHAPTER 18

GHOSTS

SEPTEMBER 1878–DECEMBER 1880

Historians have identified the period between 1861 and 1877—from the beginning of the Civil War to the end of Reconstruction—as the era that created the America we recognize today, when a continental power finally coalesced north and south, ocean to ocean. At the time of the 1878 total solar eclipse, the country was still adjusting to this new reality. Like an ungainly teenager after a growth spurt, the United States was settling into its larger, more muscular body, and it was beginning to exert its strength. Soon it would project its military might overseas, interpreting Manifest Destiny on an ever-grander scale as it grabbed possessions in the Caribbean and the Pacific, and by 1900 this brawny empire would overshadow its European rivals economically, outperforming Britain, France, and Germany in industrial production. Around that same time, America would start eclipsing the Old World in another realm: the pursuit of science—an eventuality that, a few generations earlier, many in Europe thought would never come to pass.

"It must be acknowledged that in few of the civilized nations of our time have the higher sciences made less progress than in the United States," observed Alexis de Tocqueville, the French political thinker, after his visit to America in 1831. "Many Europeans, struck by this fact, have looked upon it as a natural and inevitable result of equality; and they have thought that, if a democratic state of society and democratic institutions were ever to prevail over the whole earth, the human mind would gradually find its beacon-lights grow dim, and men would relapse into a period of darkness." Simon Newcomb did not subscribe to this view, but the American astronomer agreed that his own country faced a special challenge. "In other intellectual nations, science has a fostering mother," he maintained, "in Germany the universities, in France the government, in England the scientific societies. . . . The only one it can look to here is the educated public." In a democratic and egalitarian America, the citizenry was in charge of the nation's destiny, and therefore advancing science in the United States required convincing the populace of the value of research—that it was worth promotion and investment.

The eclipse of 1878, then, arrived at a fortuitous moment in this campaign, and its influence was far-reaching and multifaceted. "It is good for our general culture if startling astronomical events awaken an interest in the things above us," Maria Mitchell wrote, and indeed the eclipse had that effect: instilling awe about the workings of nature, educating a broad public through a flood of science news, and infusing astronomical research with a patriotic fervor as Americans cheered on their home team of observers. For the United States, a nation of strivers, the celestial event suggested a higher calling—a point made, appropriately, by a man of the cloth. "When the time of our late eclipse drew near, what a procession of arts and of instruments moved far out to where the shadow would fall!" the Reverend David Swing, an influential preacher, sermonized from the stage of the fashionable McVicker's Theater in Chicago in late 1878. "[I]n the very summer when we are lamenting most that mankind knows no pursuit except that of gold," he proclaimed, "[this] Rocky Mountain scene only faintly illustrates the intellectual activity of our era. If the passion for money is great in our day, it is also true that the intellectual power of the same period is equally colossal." Americans radiated pride when speaking of the astronomical event that their nation had hosted. As one Centennial State newspaper exulted, "Colorado has . . . furnished the grandest eclipse of the age."

The federal institutions that had sponsored expeditions claimed a share of the glory as well. The U.S. Naval Observatory preened over what its efforts had produced: an abundance of new photographs, drawings, spectroscopic data, and other observations for scientists to ponder in the years ahead. "It is thought that the results reached are important, and in many respects differed from things previously seen," Admiral John Rodgers, head of the observatory, wrote to the secretary of the Navy. "In many respects this is perhaps inferior in interest to no eclipse ever observed, since the weather was exceptionally fine, and it was attempted to take advantage of the previous experience of the world." The Naval Observatory prepared a thick volume of reports for wide distribution.

The U.S. Army Signal Service seemed to be making similar plans. General Myer assigned Cleveland Abbe to compile a report, for presumed publication, on the eclipse observations from Pikes Peak and from weather service observers elsewhere. Abbe avidly took up the task. He was eager to promulgate a theory he had devised just after totality, that those vast rays he had seen emanating from the corona were enormous streams of meteors pouring into the sun. This idea aligned with another theory some nineteenth-century astronomers embraced—that the sun's energy was generated by a constant rain of material from space—and Abbe further proposed that his meteor streams might account for periodic displays of shooting stars on earth. Abbe surmised that meteor showers occurred when our planet passed through these highways of debris. (In fact, we know now that such showers result from the earth passing through dust trails from comets, and the sun derives its energy from nuclear fusion, a process unknown in Abbe's time.)

Several months after the eclipse, in October 1878, Abbe gave his boss his lengthy report—a full 461 pages—at which point the general, perhaps still reluctant to share the spotlight with his underling, put a stop to the whole project. Only General Myer's death, two years later, would finally give Abbe the chance to publish his report, but by then Abbe had long since shared his hypothesis about the coronal streamers—and his ordeal on Pikes Peak—with other scientists and the press, and he must have been cheered by the response, which helped restore not only his reputation as an astronomer but also his nickname. "Was ever [an] astronomer in such a plight?" *The Chicago Tribune* wrote in a humorously flattering article. "Sick, near-sighted, flat on his back, trying to squint through a pair of old, defective spectacles," and yet Abbe had made what seemed a potentially important discovery. "Now if Old Probabilities, with only a bad pair of spectacles, a few obliging young men, and good Mrs. Copley, and flat on his back at that, could do so much, what might he not have done if he had had a telescope, and a tasimeter, and a galvanometer, and an integrating telescope, and could have stood on his feet?" Before long, periodicals in London were also recounting Abbe's deed. France and England had long boasted stories of eclipse adventure—Jules Janssen's balloon escape from Paris under siege, Norman Lockyer's shipwreck on the way to Sicily. Now America could claim its own tale of pluck and tribulation in the race to catch the shadow of the moon.

Scientists in the Old World offered generous praise for what the United States had accomplished. Lockyer declared in the pages of his own *Nature*, "I regard this eclipse as the most important that has been observed for many years as it throws much needed light on many points hitherto obscured in doubt." Scotland's Astronomer Royal Piazzi Smyth, in a letter to Henry Draper, extended his congratulations on the eclipse, "which American men, and American instruments, methods, & ideas, have made more peculiarly & grandly American, than any Solar Eclipse you have had in your country yet." For those scientists on the receiving end of the praise—the Americans who had gone west seeking fame and respect and societal change in the ethereal glow of the sun's corona—the event's influence would endure, both for good and for ill.

IN THE FALL OF 1878, three months after the eclipse, Maria Mitchell once again attended the annual Woman's Congress, this time in Providence. In addition to the usual crowd of activists and intellectuals, a large contingent of the general public attended. Schools cancelled classes so teachers and older students could hear the lectures. More than a thousand tickets were sold. Two years earlier, in 1876, when Mitchell had presided at the gathering in Philadelphia, she had delivered the opening remarks. Now, as a vice president of the assembly, she was advertised to speak on the final evening.

At 7:30 P.M. on Friday, October 11, the audience pressed into Low's Opera House, filling the ground floor and boxes. Up on the stage stood a blackboard. Mitchell ascended and drew three circles in chalk to represent bodies in our solar system. "In a total eclipse of the sun, the earth passes through the small dark cone of shadow which the moon throws upon the earth," she explained,  adding lines to her diagram to indicate the area in shade. "The dark shadow is never great in extent, at most but about 200 miles, but as the moon moves along it throws its shadow upon one place after another, circle of black overlying circle of black, until if we map down the footsteps of the shadow a narrow band of black seems to girdle the country." Mitchell described the path of totality of the recent eclipse, a line that crossed rugged terrain from Montana to Texas. "Looking along this dark strip on the map, each astronomer selected his bit of darkness on which to locate the light of his science," she said. "My party chose Denver."

Here she turned from professor to storyteller, relating the adventures of the Vassar College eclipse expedition. She told of adversity— the storms that plagued the region, the railroads that misplaced her luggage. "We haunted the telegraph rooms and sent imploring messages," she recalled of her desperation. She praised those who had come to her party's aid—a "friend who lived in Denver" (Alida Avery), and the Sisters of Charity. "[T]he black-robed sweet faced women came out to offer us the refreshing cup of tea and the new-made bread." And she described the flawless sky that met totality. "As the last ray of sunlight disappeared the corona burst out, all around the sun, so intensely bright near the sun that the eye could scarcely bear it."

Thomas Edison's lecture on the eclipse, in St. Louis, had been technical and dry. Maria Mitchell's approached poetry. "Looking toward the southeast we saw the black band of shadow moving from us—160 miles over the plain and toward the Indian Territory. It was not the flitting of the cloud shadow over hill and dale," she said, "it was a picture which the sun threw at our feet of the dignified march of the moon in its orbit." The astronomer finished her performance to great applause, "conclusive proof," a Chicago newspaper noted, "that Miss Mitchell had popularized science."

If at previous meetings of the Woman's Congress the press had sometimes ridiculed the speakers as grotesque harpies, the reviews now seemed unequivocally positive. "Professor Maria Mitchell of Vassar, who came to describe the solar eclipse at Denver with graphic and beautiful language, must have satisfied every man fortunately present that the highest scientific attainment is compatible with true womanliness," a Boston daily commented. "Possibly they may have felt that she gave a masculine stroke in drawing her diagrams, but would certainly concede that she proved herself perfectly feminine when she lost her trunks."

Mitchell's story of robustly feminine Vassar astronomers surely helped her cause—transforming public attitudes toward women in science. However, proponents of the broader effort to expand women's education were still haunted, as the founding dean of Bryn Mawr College later put it, "by the clanging chains of that gloomy little specter, Dr. Edward H. Clarke's *Sex in Education*." The idea had not yet been banished that education itself could coarsen and sicken American girls. By the early 1880s, though, a group of female college graduates would form an organization, the Association of Collegiate Alumnae (now the American Association of University Women), which would boast among its members a physician, Dr. Emma Culbertson, Vassar '77, who had received her medical degree after participating in Mitchell's eclipse expedition to Denver. The association would immediately take up Dr. Clarke's screed and answer it with logic by surveying more than seven hundred alumnae, inquiring extensively into their health (including menstrual health) before, during, and after college. The completed questionnaires were then delivered for analysis to an impartial outsider, the head of the Massachusetts Bureau of Statistics of Labor. His findings, issued in 1885, were clear: "The graduates, as a body, entered college in good health, passed through the course of study prescribed without material change in health, and since graduation . . . do not seem to have become unfitted to meet the responsibilities or bear their proportionate share of the burdens of life." College did not harm women, the report emphasized, and the evidence was so conclusive "that there is little need, were it within our province, for extended discussion of the subject." In other words, the case was effectively closed, and while the ghost that Dr. Clarke had conjured would linger for some time, its reign of terror finally began to lift, due—fittingly—to a scientifically minded group of women.

Overall, the 1880s would be a time of energy and optimism for women's higher education, but Mitchell herself, after a long run of good health, slowed down. Two years after her journey to Colorado, she fell seriously ill and left Vassar to convalesce with family on Nantucket. "If I am to get really well," she wrote a short time later, weak in body but strong in resolve, "may I be enabled to work for others & not for myself." She did just that, returning to Vassar to teach and fight for women's rights until she physically could do so no longer. She died at age seventy in 1889, still thirty-one years before American women were finally granted the vote.

JAMES CRAIG WATSON, less selfless, focused his energies after the eclipse of 1878 on his own renown. Famous astronomers in Germany and France sent letters praising his "brilliant discovery," and even Alexander Graham Bell—who had received a medal from Watson and the Group 25 jury at the Centennial—applauded his former judge. "Permit me to congratulate you upon the fame that has come to you as the discoverer [of Vulcan]," wrote the inventor of the telephone.

Watson leveraged his newfound prestige to further his career. The University of Wisconsin was building a grand new observatory and wooed him as its director, and despite the University of Michigan's vehement attempts to keep him in Ann Arbor, Watson defected to Madison for more money and a larger telescope. His new home state bragged of landing such a hefty prize. "[H]e will be a most valuable addition to the faculty of the university," the *Wisconsin Journal of Education* gloated, adding, in a line that would prove morbidly ironic, "He is in the very prime of life—a man of great physical and mental vigor."

During this period, however, the story of exactly what Watson had discovered in Wyoming changed. This shift occurred, at first, for a propitious reason. After the eclipse, when scientists were puzzling over why no one but Watson had seen the intra-Mercurial planet, it soon turned out that somebody had. The second charmed observer was Lewis Swift, a Rochester hardware merchant and respected amateur astronomer who had been among the throngs on Denver's Capitol Hill. Upon returning home, Swift announced that he, like Watson, had spied an unknown object that shone with a reddish light near the star Theta in Cancer. The news was exactly what Watson needed—corroboration—and since totality had occurred in Wyoming prior to Colorado, Watson retained credit for Vulcan's discovery. He had seen the planet several minutes before Swift.

The situation was not quite so simple, however. During totality, Swift had actually noted two objects; he had assumed one was Vulcan and the other was a star. Watson too had noted a second object, which he also had taken for a star; but when he reexamined the object's coordinates as marked on the cardboard circles he had attached to his telescope, he began to change his mind. "I have no doubt about being able to substantiate the discovery of one intra-Mercurial planet," he wrote to the U.S. Naval Observatory in early September, "and I am pretty sure that I shall establish the existence of two." Before long, Watson was certain. "[T]he records of my circles cannot be impeached by all the negative evidence in the world. There are no known stars in the places which they give, and hence I cannot be mistaken as to the identity of the objects which I observed." A Chicago newspaper remarked, "This Vulcan business grows interesting and confusing."

It soon grew more confusing. Lewis Swift discovered that he had miscalculated the coordinates of the objects he had seen. When he corrected the data, it became evident that his objects matched up neither with the ones that Watson had seen nor with known stars. "If the above conclusions are true," Swift wrote, then "four planets were discovered instead of two."

The strange saga of the multiplying Vulcans played out in the pages of *Nature* and other science journals, and it prompted some of Watson's colleagues to snicker behind his back. Simon Newcomb, in a letter to C. H. F. Peters in 1879, remarked that the situation seemed to mirror, in reverse, a nursery rhyme known today as "Ten Little Indians" but which, in that era, was sung at minstrel shows and went by an even more derogatory title, using an epithet for African Americans. In the rhyme, a group of boys shrinks one by one as a series of accidents befalls them. "But," Newcomb wrote, "the Vulcans have increased by one with every accident." Newcomb then supplied the start of his own ditty, penned in Peters's native German, but here translated, loosely, to retain the rhyme:

One little Vulcan tastes a scalding stew,
He rockets off his little chair and look! Now there's two.
Two little Vulcans etc.
And look, now there's three.

Peters, as Watson's longtime nemesis, read it with delight. "I should have liked to hear the whole of that poem, of which you only give me the beginning," he replied. "Here with us the poetical fountain runs very badly, but I might continue perhaps in this way," he wrote, picking up the rhyme at "three"—again mostly in German, and translated here:

Said one to the other, "Let's have a beer, and more!"
They kept on drinking merrily—and look, now there's *four*!
They wanted yet another, a *fifth* would give them breadth.
Alas, the keg is empty—they drank themselves to death.
To which may be added perhaps the lesson:
That's what always happens, the frog puffs and puffs galore;
In the end he bursts apart—no one believes him anymore.

Peters was sure that his puffed-up rival would soon be deflated and that no one would believe him anymore. Indeed, Peters intended to do the puncturing. Ever convinced of Vulcan's nonexistence, Peters attacked Watson in what was then the world's leading astronomical journal, a German publication called *Astronomische Nachrichten*. The critique was snide and damning. After deriding Lewis Swift's observations as so unreliable they were hardly worth mentioning, Peters set his sights on Watson's assertion that the locations he claimed for his supposed planets "cannot be impeached." Any measurement, after all, comes with a margin of error, so Peters carefully considered the unusual method Watson had used for noting the coordinates of his objects: the wood-and-wire contraption that a Rawlins carpenter had attached to the telescope the night before the eclipse. Peters concluded that making marks with a pencil pressed against a flexible metal pointer, "in the dim light of the total eclipse" and "with a certain hurry," would have introduced imprecision, perhaps amounting to one-seventieth of an inch. This tiny error on the cardboard, Peters showed, took in enough of the sky around Watson's stated coordinates that the position of each "planet" could actually be that of a prominent star in Cancer. "It is therefore quite apparent to every unbiassed [sic] mind, that Watson observed [these stars], nothing else," Peters asserted. He was accusing Watson of a beginner's error, mistaking fixed stars for planets.

Watson seethed. "Professor Peters' whole attack upon the integrity of my observations is not of the slightest consequence, since he has created the errors in his own brain and has then produced [them] to assail them," Watson huffed. "I do not intend to engage in any controversy about these matters and especially with a person who was, at the time of the observations, more than two thousand miles away from the place where the eclipse was observed." Rather than offer a full rebuttal of Peters's charges, Watson resolved to answer his rival in a more spectacular manner. He would prove that his planets existed. He would find them again, this time without the aid of a solar eclipse.

Since the age of Aristotle, a myth has persisted that by looking up from the bottom of a deep well or mineshaft, it is possible to see the stars at midday. This belief has endured, despite evidence to the contrary, because it possesses a whiff of plausibility. After all, the walls of a deep well or mine should block out the sun and allow light to enter only from a narrow portion of the sky. But the method does not work because what hides the stars in the day is not sunlight coming in at an angle, but the bright sky itself, which washes the stars out. Descending into a well does not change the sky's brightness.

A man of Watson's intelligence should have realized as much, but he was determined to restore his besmirched honor, and his move to Wisconsin gave him an opportunity for bold action. The university's observatory sat at the top of a south-facing hill, and he decided to use his own funds to build a small second observatory at the bottom of the hill, specially designed to look for Vulcan. The plan was to dig a cellar twenty feet deep and to install an underground tube, parallel to the earth's axis, that ran fifty-five feet from the cellar to the top of the hill. An adjustable mirror would direct light from any part of the sky down the shaft and into a telescope in the subterranean observatory.

Watson began construction in the spring of 1880. All through the summer and fall, while continuing his astronomical studies at night, he personally supervised by day the erection of his small observatory, as well as an expansion of the university's large one. By mid-November, the underground facility was almost ready for the installation of the telescope. "I am happy to say that it is nearly completed," Watson wrote to a scientist friend who had served with him as a judge at the Centennial.

Just then, the autumn weather turned. "Look out for a 'freeze up,'" a Wisconsin newspaper cautioned, but Watson continued to oversee his construction projects with single-minded determination. A witness to his behavior wrote that the professor went "out among the workmen on the Observatory without bundling up. I thought he was imprudent but did not like to speak to him about it." Sure enough, Watson caught a chill, yet he continued his dogged work. His cold lingered. Then, suddenly, his condition took a downward turn.

On Monday evening, November 22, 1880, two doctors were summoned to the house, but Watson was rapidly deteriorating, and they could not forestall the inevitable. Death arrived for the astronomer, age forty-two, at six in the morning. The official cause was inflammation of the bowels, perhaps appendicitis, but many discerned a deeper explanation—Watson's "recklessness of his own health" and "exhausted condition by overwork and exposure," brought on by his obsession with Vulcan.

IN HIS FINAL TRANSIT, Watson gained a share of what he so desperately craved: immortal fame. News of his death engendered effusive headlines—"The Most Brilliant of American Astronomers

Suddenly Cut Off," declared a Chicago paper—while dignitaries from Madison escorted his body back to his alma mater in Ann Arbor, where it lay in state in a hall draped in black, under the watch of a student honor guard. The school held an august memorial service during which former colleagues, in rhapsodic eulogies, recalled Watson as "the brightest scientific ornament of the University" and remembered "with what delight we received the news the day after the eclipse, of the triumphant success of his search." But the nagging questions about Watson's claimed discovery of Vulcan persisted, and his underground observatory remained unfinished.

The University of Wisconsin soon named Watson's successor—Edward Holden, from the U.S. Naval Observatory—and before he had even moved to Madison, Holden received a letter from C. H. F. Peters, as surly as ever despite his archrival's demise. "One thing I would beg you most earnestly: do not sit in that subterranean hole, to watch until Vulcan passes," Peters implored. "Not that I apprehend you might discover him, not out of selfish obstinacy in my opinion, I beg you,—but it might deadly ruin your health, and you better fill up the hole." Holden did not fill up the hole. He completed the underground facility and tested it, searching in daylight for the Pleiades and other bright stars. If the observatory enabled him to view these celestial objects in a blue sky, he reasoned, then he had a chance of finding Vulcan (if, that is, it existed). Almost two years after Watson's death, Holden finally announced his results. "No stars were seen at any time," he reported. "I am satisfied, therefore, that there is no use in prosecuting this particular experiment further." The subterranean observatory was then "abandoned as entirely useless," as one Wisconsin newspaper reported.

It would take almost forty years more for the Vulcan story to reach its dramatic denouement. Astronomers continued to look for the ghost planet—albeit fruitlessly and with decreasing zeal—at total solar eclipses in 1883, 1887, 1889, 1893, 1900, 1901, 1905, and, yet again, in 1908. Then, in 1915, it seemed the mystery might have been solved by none other than Albert Einstein. His new general theory of relativity stated that gravity warped the very fabric of the universe, which meant that Newton's law of gravitation broke down near massive objects like our sun. When Einstein applied his theory to the orbit of Mercury, his calculations perfectly described the planet's peculiar motion. In other words, there was no longer any need to invoke the existence of Vulcan—indeed, the planet must not exist—if Einstein was right.

In 1919, Einstein's abstruse theory was famously put to the test at yet another solar eclipse. Two teams of British astronomers, stationed astride the Atlantic—in the north of Brazil and off the west coast of Africa—took a series of photographs during five minutes of totality, then carefully studied the exposures to assess how the hidden sun had bent starlight, and thereby to determine whether this deflection matched general relativity's predictions. It did, and with that, Einstein supplanted Newton, and, as a consequence, the mythical Vulcan was soon forgotten, discarded as a relic of obsolete science.

James Craig Watson, however, had ensured that he would not be forgotten. Thanks to his assorted business enterprises, he had amassed considerable wealth for a college professor, and after his death, when his will was read, it was revealed that he had left the bulk of his estate not to his wife or mother, but to the National Academy of Sciences. ("[W]hile it demonstrates his large love of science," one newspaper commented, "[it] speaks poorly for his humanity.") Watson bequeathed the money to create a perpetual fund for two purposes: to continually track his asteroids so they should never be lost in the heavens, and to establish an astronomy prize to be awarded from time to time. It would be known as the James Craig Watson Medal, and it would bear his likeness in gold, so that long after his mortal remains had been buried at Ann Arbor's Forest Hill Cemetery, his ego would endure.

CHAPTER 19

SHADOW AND LIGHT

SEPTEMBER 1878–OCTOBER 1931

E DISON'S LIFE TOOK A VERY DIFFERENT COURSE FROM THAT of James Craig Watson's in the years immediately after the eclipse. While the astronomer's apparent coup in Wyoming eventually led to tragedy, the inventor's exploits pushed him toward one of the most celebrated feats in the history of technology, for as soon as he returned from the eclipse expedition, Edison turned his attention to the problem of electric lighting. By the time Edison took up this challenge, other men—able competitors—had for years been exploring ways to convert electricity to illumination. Edison was, in fact, late to enter the game, and it was one he was far from certain to win.

In 1878, electric lights were already being used in limited settings. Arc lamps, which shine by means of a glaringly bright spark, had been installed in factories and warehouses, as well as outdoors— notably in Paris, where they glimmered along the cosmopolitan Avenue de l'Opéra. Bringing the light indoors for domestic use, however, required a softer radiance, akin to gaslight. The solution appeared to lie in incandescence, the heating of a thin element until it glowed, but incandescent lamps had so far proved impractical; they were expensive to operate and tended to fizzle out quickly.

Solving these problems became Edison's frenzied preoccupation for several years, a complex drama involving scores of players and multiple acts that began when he returned from observing the total eclipse of the sun.

"My dear Edison," George Barker wrote upon his arrival back in the East. "I am greatly indebted to you for the Western trip." The physicist and the inventor had, during their long hours together—traveling to Wyoming, then California, then St. Louis—spoken at length about electric power and lighting, and Barker now urged Edison to join him on a scientific field trip. "I have arranged the trip to Ansonia that we spoke of, for Sunday next," he wrote. The attraction in Ansonia, Connecticut, was a new dynamo that illuminated a factory with a bank of dazzling arc lights. When Edison beheld the installation, he was transfixed and "fairly gloated over it," according to a New York journalist who had, unsurprisingly, been invited along. "[Edison] ran from the instrument to the lights, and from the lights back to the instrument. He sprawled over a table with the simplicity of a child, and made all kinds of calculations." Edison saw not just the promise of electric lighting, but how a power grid, similar to the network of pipes that supplied gas for domestic lighting, could illuminate a whole city.

A week later, Edison convinced himself that he had also cracked the problem of how to make a practical incandescent bulb, and he announced as much to the press. "I have it now!" he told the New York *Sun*. "When the brilliancy and cheapness of the lights are made known to the public—which will be in a few weeks, or just as soon as I can thoroughly protect the process—illumination by carbureted hydrogen gas will be discarded." The public response was immediate. Gas company stocks plunged. Wealthy investors hounded Edison, eager to get in on the ground floor. Before long, the Edison Electric Light Company had been incorporated with a capital stock of $300,000, and Edison was expanding his workshop to bring the new lamp to fruition.

There was, however, a problem: Edison had made his pronouncement based on sheer self-confidence. It was the same modus operandi he had employed over the summer with the aerophone, phonomotor, and tasimeter—dream up an idea, proclaim it a reality, and only then figure out how to make it work. "When do you expect to have the invention completed, Mr. Edison?" he was asked in October 1878, a month after he announced his breakthrough in electric lighting. "The substance of it is all right now," Edison replied confidently, "but there are the usual little details that must be attended to before it goes to the public." In truth, those "little details" were everything. Edison did not know how to make a practical lamp, a fact he soon came to realize.

At this time, the tasimeter, too, was revealing its impracticality. After returning from Wyoming, Edison received inquiries from amateur astronomers and professors alike who were eager to try the delicate heat measurer themselves. Uncharacteristically, Edison chose not to patent the invention—"it is of no interest to the people, but only to scientists," he explained—and he permitted two companies, one in London and one in Philadelphia, to make and sell the instrument without paying him royalties. When scientists finally had a chance to test the device, however, they discovered its deficiencies. "The movements of the needle are very erratic," one scientist complained. Another wrote that "the instrument has been generally regarded as peculiarly inconstant and unreliable in its indications." Although the tasimeter could detect minute amounts of heat, it could not dependably quantify that heat. In other words, it did not really *measure* anything; its results were irreproducible. As a newspaper remarked after the eclipse, "Edison's tasimeter was like the woman who recently cut off her husband's head because he found a fly in the butter—too sensitive by half." Scientists tended to agree, and they soon lost interest in the tasimeter. Even Edison acknowledged years later that his eclipse experiment had been of no consequence. "My apparatus was entirely too sensitive," he admitted, "and I got no results."

ALTHOUGH EDISON'S TRIP TO Wyoming had been a bust scientifically, it had been a great success in another way—as a demonstration of his mastery of public relations. Edison had generated enthusiasm about the tasimeter even before he had built it, and he sustained that interest despite the invention's failings. This was no small matter. Indeed, these same skills now proved critical to the success of his electric lamp. Having announced to the world that he had solved the problem of incandescent lighting—when he had not—and having raised money on that promise, Edison was in a fix. Weeks passed, then months, as he and his men worked incessantly. To create a viable lamp, they needed to find a filament (a term Edison coined in this context) that would glow when heated but would not melt or burn out quickly. To prevent the filament's combustion, they had to devise new methods for creating a near-perfect vacuum inside the bulb.

Meanwhile, Edison designed other elements of the system in which his bulb would operate, including improved electric generators, an electric meter, and insulated underground wires that could safely distribute electricity to homes. Still, even a year after he claimed to have solved the problem of incandescent lighting, he had not created a workable bulb, and he was forced to delay, time and again, his promised demonstration of the invention. People began to wonder if the Wizard was a sham.

Edison's relationship with the press now proved critical, for he needed to prevent the public and his investors from abandoning him. One tack he would soon take was to go into publishing himself; he would underwrite the launch of a weekly magazine, called *Science*, which would defend him in its editorials without ever mentioning his financial involvement. A more immediate concern, however, was the newspapers. Once again Edison relied on a key ally, his roommate in Rawlins, Edwin Marshall Fox of *The New York Herald*. "[K]eep yourself aloof and reserved," Fox urged, advising Edison to be more discriminating in the reporters he talked to, and in what he said. "Every word that you utter for publication should be to accord with your reputation like thunderbolts." The two men forged a mutually beneficial partnership. Edison fed Fox scoops and, occasionally, money: a $125 loan and a gift of eight shares in the Edison Electric Light Company, an obvious conflict of interest that worked in the inventor's favor. In return, Fox wrote flattering profiles, and, toward the end of December 1879, after Edison had finally devised a workable bulb—using carbonized paper as a filament—it was Fox who broke the news in a lengthy feature headlined "The Great Inventor's Triumph in Electric Illumination." Fox called the light "a little globe of sunshine, a veritable Aladdin's lamp."

Scientists, however, were not convinced that the lamp was as magical as Fox portrayed. Would it prove cost-effective and durable enough for everyday use? How did it differ from incandescent lights being developed by competitors? Fueling the skepticism was the fact, now abundantly clear, that Edison had been stringing the public along for more than a year by telling what could generously be described as embellishments, more reasonably termed lies, about his progress on the invention. Some scientists who knew Edison well, who had been with him on the American frontier for the eclipse of 1878, came to disavow his membership in their fraternity. Britain's Norman Lockyer, who after seeing Edison at work in Rawlins had praised the inventor as "no unwary experimenter," now denounced his actions as exposing an "absolute incompatibility with a truly scientific spirit," and added sharply, "Let scientific men once and for all repudiate these false and unwholesome displays of ignorance." Henry Morton, who had perched with Edison on the Union Pacific cowcatcher and had privately expressed himself "under so many obligations to your kindness," now became one of Edison's severest critics. Morton called Edison's electric light "a conspicuous failure, trumpeted as a wonderful success . . . nothing less than a fraud upon the public." Even George Barker turned traitorous at one point, publicly praising one of Edison's rivals for inventing "a lamp which surpasses, I believe, even Edison's dreams."

"Grif!" Edison shouted to his personal secretary, Stockton Griffin, on a day when the barrage of insults grew too much to bear. "When I am through with my light, when it is running all right in New York, I'm going to attend to these men," Edison said, imagining his revenge. "I'll build a monument with a base fifty feet wide. On this I'll put the names of all of these so-called scientific men, together with what they have said of my light. The column shall be capped with an ass's head, and this inscription, 'But remember, my masters, that I am but an ass; though it be not written down yet, forget not that I am an ass.'" Edison then completed the mental picture. "At night," he concluded, "I'll have the thing lighted up with electric jets."

EDISON'S CHANCE TO CONFUTE his critics arrived the following year, in the fall of 1881, when the world came to Paris to celebrate the dawn of a new age. On a triangle of land between the Champs-Élysées and the Seine, inside the gargantuan Palace of Industry, the International Exposition of Electricity was underway, showcasing myriad and novel uses for the mysterious force once caught by Franklin's kite. There were electric lathes and clocks, electric hairbrushes and stove tops, an electric piano and an electric dumbwaiter. Edison's team displayed its electrical wares upstairs in the eastern pavilion, in two opulent rooms with paintings on the walls and, on tables in the middle, his many newfangled contraptions, including the carbon telephone, the quadruplex telegraph, and the tasimeter, which still held interest for the public if not so much for astronomers.

The exposition's real attraction, however, occurred at night, when the lights came on—hundreds of them. Arc lights blazed across the hall's expanse, while mellow incandescent bulbs glowed in chandeliers and lanterns and globes. The lights represented the work of more than a dozen inventors, including Edison—who had by now switched to a more durable filament, Japanese bamboo—and in this way the Paris show provided a venue for the public to compare rival

L'EXPOSITION DE M. EDISON
PARIS. — L'EXPOSITION INTERNATIONALE D'ÉLECTRICITÉ
AU PALAIS DE L'INDUSTRIE

systems of illumination side by side. "Surely the game of 'throwing light' could hardly be played with a finer company of contestants or on a fairer field," remarked the *New-York Tribune*.

This was, indeed, a contest. Like Philadelphia's Centennial, Paris's exposition had impaneled a jury of scientists to bestow medals based on an array of criteria. "In the matter of the diminutive incandescent light alone," *The Cleveland Leader* reported, "it will be necessary to determine as to its intensity, steadiness, and color; the number of lights in a series; the tendency to become extinguished; the motive power per candle, distance of generator, thickness of wire, and most important, the cost for power."

The jury weighed and deliberated, and on October 21, 1881, distinguished guests crowded Paris's Conservatory of Music to hear who would receive the awards. It was a grand ceremony, laden with speeches and a choir's performance that "filled the hall with anthems of success for the exposition and for the future of electricity." Edison, however, was not there to learn the results. He was an ocean away, already working to commercialize his invention—tending to his new lamp factory in New Jersey and constructing a power network in New York that would soon electrify Manhattan's financial district. The news from Paris arrived by wire.

"Accept my congratulations," read the message from a judge. "You have distanced all competitors and obtained a diploma of honor, the highest award given in the exhibition." Edison's lamp had prevailed. "No person in any class in which you were an exhibitor received a like award," the cable concluded. It was signed: "George F. Barker."

IN THE YEARS THAT FOLLOWED, as his public acclaim and wealth continued to grow, Edison reconciled with America's scientific elite, if somewhat tepidly at times. In 1889, Henry Morton acknowledged of the electric light: "It is to Mr. T. A. Edison, without doubt, that we owe many of the simplifications and modifications which . . . have extended its range of use and its usefulness to a remarkable degree." George Barker, again one of Edison's staunchest allies, in 1908 pressed to have his friend elected to the elite National Academy of Sciences. (The academy would induct Edison, but not until 1927.) Yet Edison did not call himself a scientist in those later years. He came to realize that he and the academics differed fundamentally. They were in the business of discovering reality; he aimed to *create* it.

In a bigger way, Edison reflected—and helped define—what it meant to be American. Like the young country that had forced its way across a continent to build cities and railroads, and then emerged as a global power at astonishing speed, the inventor evinced an unwillingness to be held back by the past and what others said was impossible. He envisioned a bright future, and he pushed all obstacles aside to reach it.

Fifty-three years after the eclipse that inspired America, on October 18, 1931—a time when the nation needed inspiration again—Edison died, at age eighty-four, in West Orange, New Jersey, where he had moved from Menlo Park decades earlier. This newer home, a red brick mansion on an estate called Glenmont, was suited to what he had become—a captain of industry and an American icon, a man who through his more than one thousand U.S. patents was credited, more than anyone, with propelling civilization into modernity. "This electrical age is largely of his creation," *The New York Times* eulogized. President Herbert Hoover asked Americans to turn off their lamps for a synchronized minute in the dark, a reminder of how the Wizard had brought light from the gloom—something to be thankful for even in the depths of the Great Depression.

Tributes arrived from across the globe, from the likes of Albert Einstein and Pope Pius XI, while locals expressed their gratitude in person. More than forty thousand mourners—workers in overalls and mothers with children—bowed their heads as they filed past the bronze coffin, which had been placed in Edison's grand library, a high-ceilinged space paneled in dark wood. Edison had used this room as his office, and he had filled its alcoves and balconies not only with tiers of books but also with a trove of mementoes, the souvenirs of a rich life. These were the items that now surrounded his body as it lay in state: portraits of presidents and other men of fame and substance whom he had befriended; a marble statue of a winged boy holding aloft a light bulb; framed diplomas and prizes; and a grainy picture in black and white, a photograph from Wyoming Territory, taken on a July day in 1878 when the young inventor and a team of astronomers awaited the illuminating darkness.

[See caption page 144]

TENDRILS OF HISTORY

IN THE SIERRA MADRE MOUNTAINS OF SOUTHERN WYOMING, two miles west of the Continental Divide, a monument stands today to Thomas Edison. The bronze plaque, embedded in a pedestal of concrete on the shoulder of State Highway 70, overlooks rugged Battle Lake, its turquoise waters set against a slope of granite and spruce. This was the very spot where Edison went trout fishing after the eclipse of 1878 with George Barker, Major Thornburgh, and a group of men from Rawlins, and the historical marker tells a story passed down in local lore—that it was here that the Wizard of Menlo Park, in a flash of inspiration, divined his light bulb.

It is, of course, a fanciful story, one that apparently originated with Robert Galbraith, the Union Pacific railroad mechanic in whose chicken coop Edison had mounted his tasimeter. Galbraith joined Edison on the fishing trip and recalled almost a half century later, "After we had been [at Battle Lake] about three days, one morning at the breakfast table, Edison was asked by Professor Barker, 'Well, Tom, how did you rest last night?' 'Well,' he said, 'I wasn't thinking about resting. I lay and looked up at the beautiful stars and clear sky light, and I invented an incandescent electric light.'" (Later embellishments suggest that Edison had been inspired by the bamboo fibers of his fishing pole, perhaps after the pole had been broken and thrown in the campfire, which caused the wooden filaments to glow.) It is true that Edison's Wyoming adventure helped foster the creation of his bulb, in subtle ways—the trip, as friends and coworkers noted, was just the vacation he needed, leaving him relaxed and ready to take on new projects—but the claim that Edison was struck by a wilderness epiphany is not credible. It does not square with his actions or statements upon returning home.

The fable written on the roadside marker is admittedly appealing, however, for it conforms with popular notions of science—that discovery is intuitive, that insights come suddenly, that the emergence of ideas is easy to trace. In reality, science is a messy business, a point made eloquently by Samuel Pierpont Langley, the astronomer who inspired Edison to invent the tasimeter and who climbed Pikes Peak with Cleveland Abbe. A decade after the eclipse of 1878, at the end of his term as president of the American Association for the Advancement of Science (the same guild that Edison had joined in St. Louis), Langley gave a farewell address in which he reflected on the progress of science and, more provocatively, critiqued how textbooks frequently misrepresent scientific progress:

> We often hear it, for instance, likened to the march of an army towards some definite end; but this, it has seemed to me, is not the way science usually does move, but only the way it seems to move in the retrospective view of the compiler, who probably knows almost nothing of the real confusion, diversity, and retrograde motion of the individuals comprising the body, and only shows us such parts of it as he, looking backward from his present standpoint, now sees to have been in the right direction. . . . With rare exceptions, the backward steps—that is, the errors and mistakes, which count in reality for nearly half, and sometimes for more than half, the whole—are left out of scientific history; and the reader, while he knows that mistakes have been made, has no just idea how intimately error and truth are mingled in a sort of chemical union.

The accumulation of scientific knowledge does not occur in a simple, linear fashion. Doctrines embraced in one generation are jettisoned the next. Seemingly productive avenues of research abruptly dead-end. Scientific discoveries and events acclaimed in their day fade into obscurity with the passage of time.

Such is the case with the total solar eclipse of 1878, for reasons that may seem obvious today, looking back from our twenty-first-century perch. Despite James Craig Watson's vaunted discovery, there is no planet Vulcan. Edison's celebrated tasimeter has long been forgotten as an obscure flop. Maria Mitchell's all-female expedition to Denver, despite the favorable attention it received in its time, failed to transform the male-dominated world of science, a realm that still too often pushes women to its margins. And yet the eclipse of 1878 did leave indelible marks on America, just not in a manner that anyone could have predicted, for such is the nature of scientific progress: less an organized march than a series of stumbles.

IF ANYONE WHO OBSERVED the eclipse of 1878 could be said to have achieved what he most wanted out of the celestial event, it was arguably Simon Newcomb. The learned astronomer, who had so lamented America's scientific standing at the time of the centennial, expressed a much brighter outlook two decades later. "[O]ur traditional reputation has not been that of a people deeply interested in the higher branches of intellectual work," Newcomb expounded in an 1897 speech. "Men yet living can remember when in the eyes of the universal church of learning all cisatlantic countries, our own included, were *partes infidelium*"—uncultured lands. As the twentieth century neared, however, Newcomb could see how far American science had progressed, for example in astronomy. "[T]o-day our country stands second only to Germany in the number of researches being prosecuted, and second to none in the number of men who have gained the highest recognition by their labors." Modern historians generally agree that by 1900 America had matured into a scientific peer of Europe, and then would soon surpass it. Come the latter half of the twentieth century, the United States would unquestionably lead the world in science, whether measured by money spent or Nobel Prizes earned.

It would be folly to claim—and Newcomb never did—that America's soaring scientific fortunes could be attributed to three minutes of midday darkness in the summer of 1878. Even before the eclipse, the United States was fast on its way to becoming a formidable scientific power. It is fair to say, however, that the celestial event helped push the country toward that destination, and not solely because it inspired and educated a broad American public.

Newcomb, in his 1876 jeremiad critiquing the sad state of the nation's research establishment, laid out several concrete steps he believed the United States must take to address its deficiencies and become, as he put it, "the leader of the world in science at no very remote day." For one thing, he argued, the nation needed new scholarly publications to disseminate its discoveries. "Not only is our scientific literature of every kind meagre in the extreme, but the facilities for the publication of any kind are extremely restricted," he wrote. In this regard, Edison's foray into publishing—inspired, some suggest, by his meeting Norman Lockyer in Wyoming—proved hugely influential. *Science*, the magazine Edison founded in 1880, became, and remains today, the nation's premier weekly journal of scientific research, America's answer to Britain's *Nature*. Although Edison withdrew his funding after just eighteen months—"It cost me too much money to maintain," he said—the quirky inventor, by starting the magazine (which was soon resurrected by Alexander Graham Bell), gave his stodgy scientific counterparts a lasting gift.

Another need of American science circa 1876, according to Newcomb, was a means to inspire and motivate researchers by publicly honoring their accomplishments. "The precise form which such a recognition should take is comparatively unimportant, but the most natural one would seem to be that of medals or testimonials to be awarded from time to time," he wrote. Here, James Craig Watson's early death contributed to Newcomb's grand design. Watson's bequest to the National Academy of Sciences, establishing the James Craig Watson Medal, was the first donation received by that organization to create a prize, and it set a precedent. Two years later, when Henry Draper also died prematurely, his wealthy wife, Anna—who had counted the seconds during totality in Rawlins—endowed a second professional prize, this one in her husband's name. Today, the National Academy bestows more than two dozen awards to recognize and encourage scientific achievement.

Of course, journals and prizes are meaningless without an adequate corps of researchers to do the actual work, and here lay another weakness Newcomb identified in centennial America: "We are deficient in the number of men actively devoted to scientific research of the higher types." Newcomb used the word *men*, but in fact women made significant scientific contributions in those latter years of the nineteenth century. In his 1897 speech, Newcomb praised a monumental project at Harvard that was cataloguing thousands upon thousands of stars the way an entomologist might study a forest full of ants—sorting them into different types and varieties based on their appearance and behavior. It was painstaking work that required the patient analysis of fragile photographic plates of the heavens, exactly the sort of labor that Maria Mitchell, in her 1876 speech "The Need of Women in Science," had argued was suited to her sex. Indeed, Harvard gave the work to an all-female team, a key member of which had been a student of Mitchell's, and funding for the costly project came from Anna Draper, whom Mitchell had urged to employ "a corps of lady assistants" to encourage young female astronomers while honoring Henry Draper's memory. The project would be known as the Henry Draper Memorial.

The long tendrils of the eclipse of 1878, and of Maria Mitchell's quiet but pronounced influence, reach well into modern times. Vera Rubin, one of America's top astrophysicists in the late twentieth century, was born in 1928, almost forty years after Mitchell's death. Rubin attended Vassar in the 1940s and was asked in 1989 what had inspired her to pursue astronomy despite the societal bias against women in science. "I knew about Maria Mitchell, probably from some children's book," she replied. "I knew that she had taught at Vassar. So . . . it never occurred to me that I *couldn't* be an astronomer." In 2004, Rubin received, for her pioneering studies of dark matter, the James Craig Watson Medal.

IN THE NEARLY ONE AND A HALF centuries that have passed since July 29, 1878, revolutions have occurred in the science of astronomy, and much has changed in the role played by eclipses. On that very day when the moon cast darkness on the American frontier, *The Times* of London suggested that an era was soon to close. "It is, in fact, thought by many that the eclipse vein is nearly worked out." The paper's prediction proved hasty and extreme—even today some astronomers head to the ends of the earth to conduct research in the moon's fleeting shadow—but the age of large and cumbersome eclipse expeditions indeed passed long ago. As the sun gave up its more discoverable secrets in the late nineteenth and early twentieth centuries, the field became more specialized, and many studies of the solar corona can now be done at any time, without waiting for an eclipse, thanks to new telescopes and sun-gazing spacecraft. Today it is a small proportion of astronomers that seeks out eclipses, and the professionals are joined by a large company of tourists, who routinely travel by jetliner and cruise ship to reach the path of totality.

What has not changed over the centuries, however, is the emotional allure of total eclipses. These rare and unearthly events, when they pass overhead, suspend human affairs and draw people out of their quotidian existence. Beholding the corona still provokes chills and tears. Some look up and find God. Others discover a new passion for science, a desire to understand the workings of the sun and solar system. The impact of eclipses remains life-changing, as I myself can wholly attest.

———

SINCE MY FIRST EXPERIENCE of totality, in Aruba in that winter of 1998, I have, so far, traveled four more times to meet the moon's shadow, journeys that have felt like reunions with an old acquaintance who is available only at odd intervals and seemingly random locales. In August 1999, I flew to Munich, where my observing platform—like that of the U.S. Naval Observatory party in Central City, Colorado, in 1878—was the roof of a hotel, in this case a high terrace where guests sipped Champagne and nibbled on canapés while awaiting the show. When the sun at last revealed its delicate corona, a collective cheer lifted from the darkened city, like the cry that the old newspapers say rose from Denver. In November 2012, I again stood in the path of the moon's shadow, on the Queensland coast of Australia, where a mantle of clouds drew apart to offer a flawless view of totality, and third contact produced a "diamond ring," a gemlike point of sunlight set in a silver band around the blackened moon.

The weather proved far less promising—predictably so—in the Faroe Islands, in the remote North Atlantic, where in March 2015, due to a lack of hotel accommodations (as Denver and Rawlins experienced in 1878), I slept at a private home, in the bed of a displaced sixth grader, and on eclipse day I stood in the sleet on the edge of a sheep pasture and managed to glimpse the corona for a full two seconds—a half second longer than Norman Lockyer's view from Sicily in 1870. And then, in March of the following year, when I witnessed totality from a fishing village on Belitung Island, Indonesia—where roosters crowed as darkness approached, and the locals seemed as fascinated by the visitors from afar as they were by the spectacle overhead—I finally caught sight of Baily's beads, four dazzling pearls that vanished the moment I comprehended them yet still seared themselves into my memory. That eclipse proved memorable in another way, for precisely one *saros*—the eighteen-year period over which these astronomical conjunctions repeat themselves—had passed since 1998. It was, in a sense, the same eclipse I had seen in Aruba. The cycle had ended and then, like the seasons, it renewed.

Eclipses, I find, connect the present with the past like few other natural events. For me, personally, they are life milestones. Each forces me to reflect on who I was the last time I gazed at the corona. For us, collectively—as a society, a nation, a civilization—they can have the same indelible, life-affirming effect. They afford a chance not only to grasp the majesty and power of nature, but to wonder at ourselves—who we are, and who we *were* when the same shadow long ago touched this finite orb in the boundless void.

TOTAL ECLIPSE of the SUN.

Total solar eclipse of July 29, 1878, as seen from Creston, Wyoming Territory. This lithograph, published in 1882 as part of *The Trouvelot Astronomical Drawings*, was described as follows by artist E. L. Trouvelot: "The eclipse is represented as seen in a refracting telescope, having an aperture of 6⅓ inches, and as it appeared a few seconds before totality was over. . . . The two long wings seen on the east and west side of the Sun, appeared considerably larger in the sky than they are represented in the picture." [Trouvelot (1882:27)]
Courtesy of Public Library of Cincinnati and Hamilton County.

Composite sketch of the total solar eclipse, as seen from Denver's Capitol Hill, produced by local volunteer observers organized and trained by the Chicago Astronomical Society. "The ladies and gentlemen . . . seated themselves on the brow of the hill, between our telescopes and the sun," the society wrote in its eclipse report. "The accompanying diagram represents the *average* outline of the corona. . . . The position of the protuberances was on the solar limb, nearly opposite the letter A, in the diagram. We have made no attempt to represent them, or the curved form of the rays." [Chicago Astronomical Society (1878:11–12)] *Courtesy of Denver Public Library.*

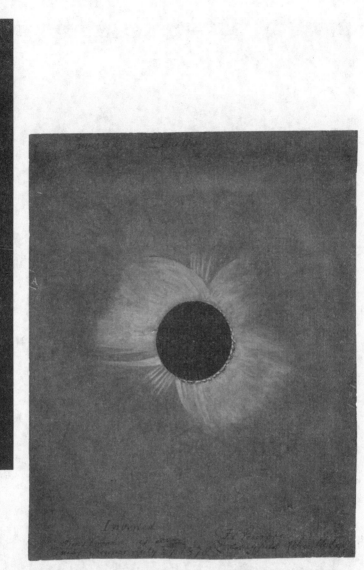

Solar corona as seen near Denver by F. C. Penrose, an "English architect, artist, and astronomer of distinction," as described by Prof. Charles A. Young of Princeton. [Young (1878b:874)] Penrose submitted this image to the U.S. Naval Observatory, which published it in its volume of reports on the eclipse of 1878. [USNO (1880:Plate 13)] *Courtesy of National Archives.*

Pencil sketch of solar corona, as observed at Separation, Wyoming Territory, by Annette Helena Watson. Her husband, James Craig Watson, had requested that she make the drawing. "I had instructed [her] beforehand how to make a proper sketch and to compare it with the sun before the total phase was over," he noted. [USNO (1880:123)] *Courtesy of National Archives.*

Watercolor of the solar corona as seen at Denver by George W. Hill, an assistant at the Nautical Almanac Office. His boss, Simon Newcomb, called Hill "the greatest master of mathematical astronomy during the last quarter of the nineteenth century." [Newcomb (1903:218–19)] Hill apologized for imprecision in his eclipse painting due to "my want of familiarity with the handling of artist's materials." [USNO (1880:221)] *Courtesy of National Archives.*

Total Solar Eclipse of July 29. 1878. As seen from Denver, Colorado.

Photograph taken at the end of totality (i.e. third contact) by the party of William Harkness at Creston, Wyoming Territory. "Before the exposure ... was terminated the sun burst out," Harkness wrote of this image. "The objective [lens] was then instantly covered, but the flash of sunshine left its mark upon the sensitive film in the shape of a beautiful set of interference rings, produced by reflection of the light at the back surface of the plate." [USNO (1880:52)] *Courtesy of National Archives.*

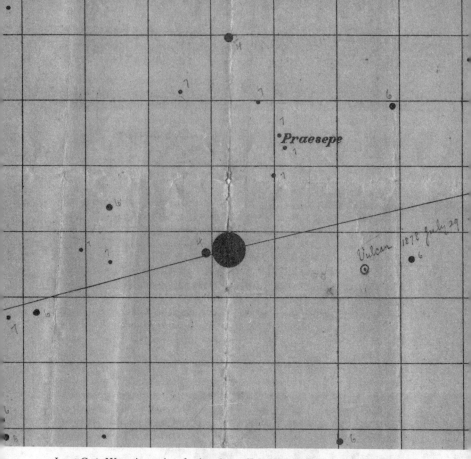

James Craig Watson's star chart for the eclipse of July 29, 1878. To the right of the large black circle, which represents the overlapping sun and moon, Watson has marked in pencil two unknown objects, the closer one labeled "Vulcan." Watson described his discovery: "Between the sun and Θ Cancri, and a little to the south, I saw a ruddy star.... [I]t did not exhibit any elongation, such as might be expected if it were a comet in that position." [USNO (1880:119)] *Courtesy of Bentley Historical Library, University of Michigan.*

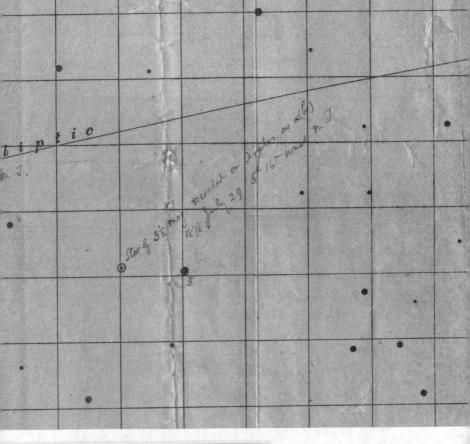

Princeton College eclipse party near Cherry Creek, Denver. Prof. Charles A. Young, the group's leader, sits second from the right. Astronomer Arthur Cowper Ranyard, a British guest at the American camp, sits on the far right. *Courtesy of John G. Wolbach Library, Harvard College Observatory.*

*U. S. Naval Observatory Eclipse Party,
Central City, Colorado; 1878. July 29.*

U.S. Naval Observatory party preparing to observe the eclipse atop the Teller House Hotel in Central City, Colorado. "This roof being nearly flat was extremely convenient for the setting up of such of our instruments as were mounted on tripods," wrote the team's leader, Edward S. Holden, "and the unused and solid chimneys served as admirable piers for the other instruments." [USNO (1880:147)] *Courtesy of U.S. Naval Observatory.*

Stereograph card of Vassar College eclipse party, Denver. Cora Harrison sits in the foreground, Maria Mitchell in the middle ground on the far left, and her sister Phebe Mitchell Kendall in the distance behind her. "The only value to the picture," Mitchell commented, "is the record that it preserves of the parallelism of the three telescopes. You would say it was stiff and unnatural, did you not know that it was the ordering of Nature herself—they all point to the centre of the solar system." [Kendall (1896:230)] *Courtesy of Archives and Special Collections Library, Vassar College Libraries.*

UNISON

Five years after this book was first published, on the weekend before Thanksgiving in 2022, a strange conjunction of light and darkness reminded me of the power of eclipses—and people—to shape lives and history. I had flown to New York for the first time since COVID-19 brought the world to a halt. In pre-pandemic days, trips to Manhattan had been an annual pilgrimage for me, a chance to enjoy the offerings of Broadway, but the virus then shuttered the theaters, and even after they reopened, I remained reluctant to brave the crowds. Now, however, I had been lured back.

On that Friday afternoon, I made my way to Times Square, to an eleventh-floor rehearsal studio where I joined a small audience for the premiere of a theatrical piece in the early stages of development. Twenty-five performers soon arrayed themselves in curving rows before music stands. They began to sing. The scene was the Centennial Exhibition in Philadelphia, and these characters—Thomas Edison, Maria Mitchell, James Craig Watson, and others—were performing the opening number of the musical *American Eclipse*, an operatic retelling of the great astronomical event of 1878. In the weeks prior, I had anxiously wondered whether my book would translate to this medium, but I need not have worried. At totality, when the chorus gazed skyward in song, I had chills. What better way to capture the beauty and awe of an eclipse than through an ensemble of voices—through harmony?

The weekend that started so joyfully, however, ended in unexpected sorrow. Two days later, on Sunday, I discovered that I had lost a friend and mentor. "Jay Pasachoff, Who Pursued Eclipses Across the Globe, Dies at 79," read the headline in *The New York Times*. "An astronomer at Williams College, he probably saw more solar eclipses than any other human in history." I was stunned. A mere month before my trip to New York, I had excitedly written to Jay and his wife about the fledgling musical. "[T]his is fabulous news," he gushed. "Naomi and I are off to Helsinki on Friday for the Oct 25 partial solar eclipse. Then we have plans for the 2023 April total eclipse in Australia." He added, "Join us"—and I had intended to. I had no idea then that Jay would soon be in hospice care with cancer.

Jay's life revolved around alignments of the sun and moon. He studied eclipses, traveled to them, extolled their awe-inspiring beauty to most everyone he met. His enthusiasm had spurred me to see my first total eclipse. To my great regret, I never had the opportunity to share an eclipse side-by-side with the great Jay Pasachoff, but when I learned of his death, I reminded myself of the legacy he left. Indeed, I and millions of others *had* stood in the moon's shadow with Jay a few years earlier, during an eclipse in its own way as noteworthy as the one of 1878.

AS MOST ANYONE WHO was in the United States at the time will recall, on August 21, 2017, the country experienced a coast-to-coast total eclipse, its first in ninety-nine years. In the runup to that day, Jay touted the event as the "first all-American total solar eclipse" because the moon's shadow would, for the first time since the country's founding, make landfall only in the United States. The projected path draped like a sash from Oregon to South Carolina.

During that spring of 2017, however, relatively few Americans were aware of what was coming. The nation was focused on other matters. It was the first year of the presidency of Donald Trump, when protests and outrage frayed the country. Partisan and cultural divides deepened: red versus blue, urban versus rural. That August—one week before the moon's shadow was set to arrive on American shores—darkness fell on Charlottesville, Virginia, where a white supremacist rally met counter protestors in a deadly clash that epitomized the country's unraveling. In such fraught times, I wondered, did a total solar eclipse still have the power to elevate and unite an entire nation, as one did little more than a decade after the Civil War?

Jay Pasachoff had no doubt that it could. An irrepressible cheerleader, he proselytized in the weeks and months before the eclipse. "[W]e're trying to persuade all 300 million Americans . . . to travel into the path of totality," he said, with only a bit of exaggeration. He repeated his message in press conferences, essays, and talks at elementary schools. "Take a kid to the eclipse," he told *The New York Times*, then elaborated in *The Washington Post*: "[T]he most important scientific outcome from this year's eclipse may be . . . inspiring a 7- or 8-year-old girl or boy somewhere to enter a career of science, perhaps even leading to a fantastically wonderful discovery 20 or 30 years from now."

Other astronomers and NASA scientists made similar appeals, as did an army of librarians and educators, but Jay was the movement's spiritual leader. He encouraged Americans to choose awe and wonder over rancor and division, and I was one of his acolytes. That summer of the eclipse, I felt like a traveling preacher. As I gave talks promoting this book from Boston to Atlanta to Seattle, I spread word of the great day that was coming. When August 21 finally arrived, it was clear that our public education campaign had succeeded wildly. From the Pacific to the Atlantic, America's focus turned outward—skyward—as myriad individuals coalesced into unified masses for this cosmic moment.

When the lunar shadow reached Southern Illinois University, fourteen thousand voices rose as one from the school's Saluki Stadium. "It just shows us how powerful we can be when we all come together, even with everything that's going on," a man in the crowd told NBC News. "You've got all different types of faces and people out here that came together for this moment. It is incredible." In Oakland, New Jersey—where townspeople gathered at the library to watch a partial eclipse but found there were not enough solar glasses for everyone to observe safely—those who had glasses shared with those who did not. "Given the experiences we've had around the country lately," a woman told the local paper, "it's good to see everyone coming together and making it work." At an enormous gathering called SolarFest in Oregon's high desert, the diverse thousands who came from all over—represented by a rainbow of pushpins on an oversized map—proved so polite and cooperative that they left almost no litter when they vacated the fairgrounds. "[I]t's immaculate," said one of the organizers, stunned.

The scene repeated itself all across the country, in parks and city streets, on mountaintops and beaches. Individuals became communities. Strangers were no longer strangers. Hardened people cried, hugged, fell reverentially silent. In this age of polarized politics, siloed entertainment, and individualized news feeds, the eclipse offered a precious shared experience—one that lifted and joined rather than debased and divided. *The Atlanta Journal-Constitution* summarized the day in a headline: "Nation united in awe."

Studies have since backed that assessment. A survey by researchers at the University of Michigan estimated that more than 150 million American adults observed the 2017 eclipse directly, while another 60 million watched it on TV or the internet. "This is a level of exposure that dwarfs the viewership of Super Bowl games and ranks among the most viewed events in American history," the study concluded. Other scientists—at the University of California, Irvine—analyzed millions of messages sent on Twitter around the time of the 2017 eclipse and found that those posted from within the path of totality and on the day of the eclipse "exhibited more awe and expressed less self-focused and more prosocial, affiliative, humble, and collective language." These findings, the team maintained, revealed the psychological impact of the 2017 solar eclipse. "Just as the moon aligned with the sun up in the heavens, people down on earth aligned with each other in awe of this spectacular celestial event."

This is what gave me solace when I learned of Jay Pasachoff's death—the many lives he touched in ways direct and indirect. Without his evangelism, and the evangelism of those he inspired, America's day of unification in 2017 would not have been so powerful. I know this for a fact. In a TEDx Talk I gave that summer, I urged the audience to follow the advice that Jay had given me in 1994. "Before you die," I implored, echoing his very words, "you owe it to yourself to experience a total solar eclipse." The talk reached a million online views before the eclipse and brought a flood of grateful messages after. "Your video was instrumental in getting 11 people from 4 families in our group out to Durkee, OR, to witness the eclipse" read an email from a father in Seattle. "It was one of the most incredible events I have ever witnessed. . . . In fact, my daughter (14) came home and now isn't sure if she wants to go to college for meteorology or astronomy." The eclipse had touched a new generation, as Jay had hoped. His influence rippled outward.

I NOW FIND MYSELF in Australia, having come here to witness another total eclipse, my eighth. During my time in the Southern Hemisphere, the reports that have reached me from America—news of mass shootings and noxious political fights over gender identity—remind me that my homeland is far from unified. Still, I remain bullish on the transcendent power of eclipses to bring people together.

Three days ago, on a peninsula that juts into the Indian Ocean in the remote northwest of Australia, I climbed a hillside that a local woman had recommended for watching the eclipse. She invited me to join her there with her friends. The terrain was treeless—sparse scrub in dry soil—and offered a sweeping view in all directions. To the east lay the small town of Exmouth. Beyond, on whitecapped turquoise waters, a cruise ship carried its haul of American tourists. Toward the south, in a makeshift campground, hundreds of eclipse chasers from across the globe (South Korea, the Netherlands, Canada) had erected tents and telescopes.

As the moon passed before the sun and the late-morning light dimmed, we all awaited the moment of full shadow. I thought of Jay Pasachoff and his plans to be here, and since he could not, I stepped in to mentor the novices who watched nervously through eclipse glasses. I advised them not to look with the naked eye until I gave the signal. "Almost," I said as the crescent sun shrank to a sliver. "Not quite yet." Then all went dark. "All right. Glasses off!"

My small group on the arid hillside screamed, and our voices merged with thousands more from town, from the campground, from the cruise ship. In my mind, our living cheers joined with those of the past, and I heard the refrain of those singing at totality in 1878, at the musical presentation in New York. The corona glowed with its unearthly light. "I don't think anything can describe that feeling," the woman who invited me to the hillside said when the sun reappeared. She looked dazed. "I felt like we were in a different world." She hugged me. "Thank you."

Talk then turned, as it always does, to the next eclipse—the next opportunity to see the solar system revealed. I explained that in less than twelve months—on April 8, 2024—the moon's shadow would revisit North America, tracing a line from Mexico to Canada across my own fractured country, which needed all the healing energy it could muster. To my new Australian friends, with whom I was now bonded by what we had just experienced, I said simply, "Come."

—Perth, Australia, April 23, 2023

NOTES ON SOURCES

I HAVE BUILT THIS BOOK ON A FOUNDATION LAID BY DOZENS of prior writers—historians, biographers, scientists, journalists—who investigated and interpreted the events and themes so central to the tale of the 1878 eclipse. Their insights proved invaluable as I sought to understand the personalities and actions of people long deceased, and as I aimed to grasp the cultural context in which these individuals lived. Please see my bibliography for a list of the many published works I consulted, but here, before I provide detailed citations largely to primary source material, I would like to highlight a few of those broader works that proved especially influential.

Anyone interested in America's intellectual maturation in the nineteenth century and the country's emergence as a scientific power should consider the following books to be essential reading: Robert V. Bruce's *The Launching of Modern American Science, 1846–1876*; Daniel J. Kevles's *The Physicists*; and Louis Menand's *The Metaphysical Club*. Through my years of writing, as I often struggled to navigate a vast sea of information, these three incisive works served as lodestars. Howard S. Miller's *Dollars for Research: Science and Its Patrons in Nineteenth-Century America* also proved a trusty guide, enlightening me about the practical steps America took to establish its scientific infrastructure in that early era. Among the books on astronomy that steered me in the right direction were Steven J. Dick's *Sky and Ocean Joined*, a comprehensive history of the U.S. Naval Observatory, and *Empire and the Sun*, by Alex Soojung-Kim

Pang, an examination of Britain's elaborate eclipse expeditions during the Victorian age. Richard Baum and William Sheehan's *In Search of Planet Vulcan* engagingly introduced me to the colorful tale of the planet that never was. Thomas Levenson's more recent *The Hunt for Vulcan*, another valuable work on the subject, approaches the story from a philosophical perspective, using the tale to explore how science progresses—and why it often goes astray.

Many fine books and scholarly articles have been written about Maria Mitchell. Of those, I am partial to Margaret Moore Booker's *Among the Stars*, a sweeping and highly readable biography. Although Mitchell never wrote an autobiography, her sister Phebe Mitchell Kendall compiled her writings in the indispensable *Maria Mitchell: Life, Letters, and Journals*. To understand the long career and deep influence of Simon Newcomb, I relied on Albert E. Moyer's *A Scientist's Voice in American Culture*. In my exploration of Thomas Edison's remarkable life, I profited from a large number of excellent books—too many to list here—as well as several journal articles that strongly shaped my writing: Norman R. Speiden's "Thomas A. Edison: Sketch of Activities, 1874–1881," David A. Hounshell's "Edison and the Pure Science Ideal in 19th-Century America," and John A. Eddy's "Edison the Scientist." (See bibliography for citations.) Another article by Eddy, "The Great Eclipse of 1878," provided the initial spark that set me on this book-writing journey, and Steve Ruskin's "'Among the Favored Mortals of Earth': The Press, State Pride, and the Eclipse of 1878" inspired me to explore the cultural aspects of this scientific tale. Without the guidance of these works—and many more—I would never have known where to begin in telling this story.

In the end, however, I have relied almost exclusively on original sources, and it is largely these items that I cite below. Many of the archival documents have not previously been published and can be found only by visiting the institutions listed. A welcome exception is the great accumulation of material compiled and organized by the Thomas A. Edison Papers at Rutgers University. How anyone wrote competently about the legendary inventor before the existence of this herculean project, I cannot fathom. Director Paul Israel and his staff continue to comb through millions of pages of Edisonia, sorting and analyzing, annotating and publishing. Where I refer, below, to items included in the Edison Papers project—whether in its books or its online digital edition—I have followed the recommended acronyms and citation format.

Many of those who participated in the tales told in this book reminisced about the events later in life, and while these recollections proved useful, they were prone to error. Stories, especially of the Old West, tend to grow taller with age. Time blurs details and promotes embellishment. In reconstructing the narrative of this book, therefore, I have relied as much as possible on contemporaneous documentation: scientific notes, diary entries, newspaper articles, letters, and telegrams. When quoting from these documents, I have preserved the original spelling and punctuation, with a few exceptions: where a handwritten item emphasizes a word or phrase through underscoring, I use italics, and where I quote what someone *said*—for instance, Mitchell's lectures, Edison's interviews, "Big Nose George" Parrott's dying plea—I have given myself license to correct antiquated spelling and punctuation to reflect how these spoken words would be transcribed today.

THE FOLLOWING ABBREVIATIONS ARE USED FOR
ORGANIZATIONS AND GOVERNMENT INSTITUTIONS:

AAAS – American Association for the Advancement of Science
AAW – Association for the Advancement of Woman/Women
NAO – Nautical Almanac Office
USASC – United States Army Signal Corps
USNO – United States Naval Observatory

THE FOLLOWING ABBREVIATIONS ARE USED FOR
ARCHIVAL COLLECTIONS:

AGBFP – Alexander Graham Bell Family Papers, Manuscript
Division, Library of Congress, Washington, D.C.

AHP – Asaph Hall Papers, Manuscript Division, Library of Congress, Washington, D.C.

AJMP – Albert James Myer Papers (microfilm from U.S. Army Signal Corps Museum), Manuscript Division, Library of Congress, Washington, D.C.

ANMP – Anita Newcomb McGee Papers, Manuscript Division, Library of Congress, Washington, D.C.

ASDR – Astrophysical Sciences Department Records (AC157), Princeton University Archives, Department of Rare Books and Special Collections, Princeton University Library, Princeton, N.J.

BMCL – Special Collections, Bryn Mawr College Library, Bryn Mawr, Pa.

CAP-JHU – Cleveland Abbe Papers Ms. 60, Special Collections, Milton S. Eisenhower Library, The Johns Hopkins University, Baltimore, Md.

CAP-LOC – Cleveland Abbe Papers, Manuscript Division, Library of Congress, Washington, D.C.

CAYP – Charles Augustus Young Papers, Rauner Special Collections Library, Dartmouth College, Hanover, N.H.

CCM – Carbon County Museum, Rawlins, Wyo.

CHFPP – Christian H. F. Peters Papers, Hamilton College Archives, Clinton, N.Y.

CWHDP – Caroline Wells Healey Dall Papers, Microfilm Edition, Massachusetts Historical Society, Boston, Mass.

DCIS – Delaware County Institute of Science, Media, Pa.

DPLWHC – Denver Public Library, Western History Collection, Denver, Colo.

EFP – Evans Family Papers, 1866–1918, Special Collections Library, University of Michigan Library, Ann Arbor, Mich.

GADRUWM – General Astronomy Department Records, University of Wisconsin–Madison Archives, Madison, Wisc.

GFBP – George Frederick Barker Papers, University Archives and Records Center, University of Pennsylvania, Philadelphia, Pa.

HFP – Howe Family Papers, Houghton Library, Harvard University, Cambridge, Mass.

HHJP – Helen Hunt Jackson Papers, Colorado College Special
Collections, Colorado Springs, Colo.

HMAPDP – Henry and Mary Anna Palmer Draper Papers, Manuscripts and Archives Division, The New York Public
Library, New York, N.Y.

HSPP – Henry S. Pritchett Papers, Manuscript Division,
Library of Congress, Washington, D.C.

JCWP – James Craig Watson Papers, Bentley Historical Library,
University of Michigan, Ann Arbor, Mich.

JEKP – James E. Keeler Papers, Special Collections and
Archives, University Library, University of California,
Santa Cruz, Calif.

JWDFP – John William Draper Family Papers, Manuscript Division, Library of Congress, Washington; D.C.

LFP – Lehmann Family Papers, Manuscripts Division,
Department of Rare Books and Special Collections,
Princeton University Library, Princeton, N.J.

LOR – Lick Observatory Records: Correspondence, University
Archives, University Library, University of California,
Santa Cruz, Calif.

MLTP – Mabel Loomis Todd Papers, Manuscripts and
Archives, Yale University Library, New Haven, Conn.

MMM – Mitchell Memorabilia Microfilm, Archives, The Nantucket Maria Mitchell Association, Nantucket, Mass.

MMP – Maria Mitchell Papers, Archives and Special Collections Library, Vassar College Libraries, Poughkeepsie,
N.Y.

NAS – Archives, National Academy of Sciences, Washington,
D.C.

NEWCR – New England Women's Club Records (microfilm),
Schlesinger Library, Radcliffe Institute for Advanced
Study, Harvard University, Cambridge, Mass.

OCMP – Othniel Charles Marsh Papers, Manuscripts and
Archives, Yale University Library, New Haven, Conn.

PSIT – Presidents of Stevens Institute of Technology Collection, Special Collections and Archives, S. C. Williams
Library, Stevens Institute of Technology, Hoboken, N.J.

SBAP – Susan B. Anthony Papers, Manuscript Division, Library of Congress, Washington, D.C.

SIA – Smithsonian Institution Archives, Washington, D.C.

SJNL – Sir Joseph Norman Lockyer Correspondence and Research Papers, Special Collections, University of Exeter, Exeter, Devon, U.K.

SNP – Simon Newcomb Papers, Manuscript Division, Library of Congress, Washington, D.C.

SPLP – S. P. Langley Papers, Bentley Historical Library, University of Michigan, Ann Arbor, Mich.

TAEB – *The Papers of Thomas A. Edison*, multi-volume book

TAEC – Thomas A. Edison Collection, Benson Ford Research Center, The Henry Ford, Dearborn, Mich.

TAED – Thomas A. Edison Papers Digital Edition (http://edison.rutgers.edu/digital.htm)

USACC – Records of U.S. Army Continental Commands (Record Group 393), National Archives, Washington, D.C.

USHR – Records of the U.S. House of Representatives (Record Group 233), National Archives, Washington, D.C.

USNAE – Photographs and Drawings Relating to U.S. Navy Astronomical Expeditions, 1878–1905 (Record Group 78-AE), National Archives, College Park, Md.

USNO-LOC – United States Naval Observatory Records, Manuscript Division, Library of Congress, Washington, D.C.

USNO-NA – Records of the U.S. Naval Observatory (Record Group 78), National Archives, Washington, D.C.

VCA – Vassar College Archives, Archives and Special Collections Library, Vassar College Libraries, Poughkeepsie, N.Y.

WB – Records of the Weather Bureau (Record Group 27), National Archives, College Park, Md.

WSA – Wyoming State Archives, Cheyenne, Wyo.

PROLOGUE: SHALL THE SUN BE DARKENED

1 **"That was the coldest weather":** All quotations and recollections in this paragraph come from *The Dallas Morning News*, Feb. 2, 1930, Automobiles and Markets Section, p. 7.

1 **Predictions of the world's imminent demise:** Gaustad (1974), Numbers and Butler (1987), Weber (1987), Butler (1991), Kyle (1998).

1 **"the earth will be dashed to pieces":** Bliss (1853:172).

1 **"will destroy the bodies":** Bliss (1853:171).

2 **"The trump of God":** D. L. Moody (1877:532).

2 **"in the clouds of heaven":** Matthew 24:30, King James Bible.

2 **"shall the sun be darkened":** Matthew 24:29, King James Bible.

2 **recently transformed from open range to farmland:** Byrd (1879:24). See also Block (1970), Anonymous (1892).

2 **"can't see to can't see":** Sitton and Conrad (2005:16). For more on the harsh lives of Texas freedmen, see Smallwood (1981), A. Barr (1996).

2 **What had motivated one Ephraim Miller:** Details of Miller's background and actions derive from an article in *The Dallas Daily Herald*, as reprinted in the *Denison Daily News* (Denison, Texas), Aug. 2, 1878, p. 1. This report appears to be the most authoritative extant account of the murder-suicide.

3 **"Exodusters":** Painter (1977).

3 **leaves bizarrely turned to crescents:** T. J. Griffiths recalled that "the shadow of each leaf assumed the crescent form" [*Dallas Morning News*, Jan. 31, 1925, p. 14], but it is really the spots of light *between* the leaves that form crescents as a solar eclipse approaches its total phase.

3 **croaking of frogs:** *Dallas Morning News*, July 29, 1928, Feature Section, p. 1.

3 **bats flying aberrantly in the afternoon:** *Dallas Morning News*, April 13, 1949, Section Two, p. 2.

3 **Fireflies winked on:** *Galveston Daily News*, July 30, 1878, p. 1.

3 **A star suddenly materialized, then two:** *Fort Worth Daily Democrat*, July 30, 1878, p. 4; *Galveston Daily News*, Aug. 1, 1878, p. 4.

3 **The air stopped moving. The birds ceased their chatter:** C. Abbe (1881:90).

3 **ripples of light:** Waldo (1879:30).

3 **Fear swept over the fields:** *Galveston Daily News*, July 30, 1878, p. 1; *Galveston Daily News*, Aug. 3, 1878, p. 4; *Tri-weekly Herald* (Marshall, Texas), Aug. 8, 1878, p. 1; *Dallas Morning News*, Jan. 20, 1925, p. 14; *Dallas Morning News*, April 23, 1949, Section Two, p. 2.

3 **A man fell to his knees:** *Dallas Morning News*, Jan. 11, 1925, p. 2.

3 **Others fled toward church:** *Dallas Morning News*, July 8, 1928, Feature Section, p. 2.

4 **"so sound asleep":** *Denison Daily News* (Denison, Texas), Aug. 2, 1878, p. 1, provides all details in this and the following paragraph.

4 **"zealous to make peace":** Herodotus ([1920]1990:91–93). For more on this ancient eclipse and attempts to date it, see Stephenson (1997:342–44); Littmann, Espenak, and Willcox (2008:49–50); Stanley (2012).

4 **"began to waste away":** Noble (2009:299). More at Todd (1894:101–2).

4 **emboldened a Native American uprising:** Drake (1858:91), Edmunds (1983:48–49).

5 **"I shall only say":** Cooper (1869:359).

CHAPTER 1: REIGN OF SHODDY

9 **MONDAY, JUNE 26:** Although James Craig Watson could not recall the precise day on which he and Sir William Thomson had tested Alexander Graham

Bell's telephone [see *Nature* 19 (475), Dec. 5, 1878, pp. 95–96], the date was June 26, 1876, as confirmed by a telegram Willie Hubbard sent to Bell that evening [AGBFP Box 321]. Bruce (1973:198) also cites this telegram.

9   **unemployment ran high among tradesmen:** *New-York Times*, March 2, 1876, p. 1.

9   **the Centennial Exhibition was a grand world's fair:** For general background on the Centennial, see United States Centennial Commission (1876); Gilmore (1876); Howells (1876); "Characteristics of the International Fair" [five-part series], *Atlantic Monthly* 38, July–Oct. and Dec. 1876; J. M. Wilson (1880); Beers (1982).

10   **aroma of tar:** Howells (1876:102).

10   **twelve hundred pounds of ice:** *Boston Daily Advertiser*, June 16, 1876, p. 4.

10   **narrow-gauge railroad:** Gilmore (1876:4); *Boston Daily Advertiser*, June 16, 1876, p. 4.

11   **colossal steam engine:** For information on the Corliss steam engine and the machinery it powered, see Rideing (1876), Gilmore (1876:8).

11   **"Dishonestly if we can":** *New-York Tribune*, Sept. 27, 1871, p. 6.

11   **another pejorative: the Reign of Shoddy:** A. K. McClure (1905:244–54) uses this pejorative to refer largely to the period from the Civil War to the Panic of 1873, but the term was still in use at the time of the Centennial. See, e.g., *New York Herald*, July 8, 1876, p. 6.

11   **"The first day crowds":** *Harper's Weekly*, July 15, 1876, p. 579.

12   **foghorn that signaled closing:** J. M. Wilson (1880:cxli).

12   **Watson seemed innately competitive:** Comstock (1895:53).

12   **"seemed to me to transcend":** Charles Young to George Comstock, Aug. 22, 1887 [GADRUWM Series 7/4/2 Box 4].

12   **"does not shrink from adorning himself":** Franz Brünnow to Edward P. Evans, July 28, 1864 [EFP].

12   **"[A]ll I can say is":** Watson to Cleveland Abbe, April 28, 1864 [CAP-LOC Box 2]. For more on the charges of plagiarism, see additional letters from and to Watson, De Volson Wood, and Asaph Hall dated between January and August 1864 [CAP-LOC Box 2]. See also Simon Newcomb to Abbe, March 11 and 16, 1867 [CAP-LOC Box 3], and Abbe to Newcomb, March 15 and 17, 1867 [SNP Box 14]. For a brief summary of the affair, see Rufus (1951:448).

12   **impoverished childhood:** Comstock (1895:46).

13   **including a student who had died:** Rufus (1951:453); *Chronicle* (Univ. of Michigan), April 2, 1881, p. 187.

13   **earned the nickname "Tubby":** *Chicago Times*, Aug. 6, 1878, p. 5; *Cincinnati Commercial*, Aug. 16, 1878, Supplement, p. 1; *Chronicle* (Univ. of Michigan), April 5, 1879, p. 177; *Chronicle* (Univ. of Michigan), May 31, 1879, p. 253; Curtis (1938:306).

13   **the jurors of Group 25:** Walker (1880:320).

13   **On the previous day, June 25:** The preliminary test of Bell's telephone on Sunday, June 25, 1876, was the more famous of the two trials at the Centennial. Alexander Graham Bell was in attendance on that day, as were many luminaries, including George Barker, Henry Draper, and most notably the emperor of Brazil, Dom Pedro. See Bruce (1973:193–98).

13   **Watson clutched a copy:** Thomson's recollections of the telephone test are in *Nature* 14 (359), Sept. 14, 1876, p. 427. Watson, in *Nature* 19 (475), Dec. 5, 1878, p. 95, noted that the items he read down the wire came from the *New-York Tribune*.

Indeed, the front page of the *Tribune* of June 26, 1876, contains phrases almost identical to those recalled by both men. The wording I use for what Watson read, and what Thomson heard, comes from the official Group 25 Report on Awards dated July 8, 1876 [TAED X031C; also in Walker (1880:452–53)].

14 **"The results convinced both of us":** *Nature* 19 (475), Dec. 5, 1878, p. 95.

14 **near the Gatling gun display:** Gilmore (1876:31–32).

14 **"It is very much like holding":** *Newark Daily Advertiser*, Sept. 17, 1875, p. 2; also reprinted in pamphlet *The Edison Electrical Pen and Duplicating Press*, p. 15 [TAED D7607J (image 9)].

15 **"The simplicity of the whole apparatus":** Walker (1880:517). See also draft Group 25 Report on Awards dated June 20, 1876, which indicates the portions of Edison's advertising pamphlet to be quoted [TAED X031A].

15 **"[A] flash of light":** TAEB vol. 1, pp. 637–38.

16 **"a very important step in land-telegraphy":** Walker (1880:452). Also see TAEB vol. 3, pp. 54–58.

16 **"Up to this very day":** Watson (1877:34–35).

16 **"knackiness":** Leng (1877:29).

16 **Foreigners scoffed:** Leng (1877:31–32).

16 **The march proved a disappointment:** J. M. Wilson (1880:cxiv); *Atlantic Monthly* 38 (225), July 1876, p. 122.

17 **"rather meagre display":** Wharton to Simon Newcomb, Feb. 17, 1876 [SNP Box 57]. The Centennial did showcase some scientific achievement by American institutions. The Smithsonian, USNO, and USASC mounted displays.

17 **"think we are a mere nation":** Newcomb to O. C. Marsh, Jan. 27 [1887] [OCMP].

17 **engaged in public self-flagellation:** Bruce (1987:342).

17 **"period of apparent intellectual darkness":** Newcomb (1876:97).

17 **Before stopping at the Centennial:** Randel (1970:91) writes that it is unclear, based on Huxley's datebooks, whether the British scientist actually attended the Centennial, but the American naturalist Graceanna Lewis claimed that Huxley did visit the exhibition. See *Woman's Journal*, Nov. 4, 1876, p. 353; *Lectures on Zoology, Miss Graceanna Lewis* (pamphlet) [DCIS].

17 **"I cannot say that I am":** *Nature* 14 (364), Oct. 19, 1876, p. 550. See more at J. V. Jensen (1993).

18 **"Our American *confrères*":** *Nature* 14 (349), July 6, 1876, p. 210.

CHAPTER 2: PROFESSOR OF QUADRUPLICITY

19 **"Menlo Park, Middlesex Co.":** Edison to Fred W. Royce, June 10, 1876 [Simonds (1934:101)]. *Louisville Times*, Oct. 21, 1931, p. 2, includes a full transcription of this letter. Josephson (1959:134) quotes from the same letter, but his wording differs.

20 **"in size and external appearance":** *Scientific American*, July 13, 1878, p. 17.

20 **"dragon's blood":** *Boston Evening Transcript*, May 23, 1878, p. 4 [TAED SM029086b].

20 **on a clear day, one could make out the towers:** *Newark Morning Register*, May 3, 1878, p. 1 [TAED MBSB10575X]; *Newark Daily Journal*, May 3, 1878, p. 3 [TAED MBSB10572X]. See also articles (cited in chapter 9) by Boston journalists who gazed at the bridge through Edison's borrowed telescope during their visit to Menlo Park on May 22, 1878.

20 **bronzes and busts and a pianoforte:** *Daily Evening Traveller* (Boston), May 23, 1878, p. 2 [TAED SM029106a]; Mary Edison insurance policy, April 7, 1876 [TAED D7606A; also in TAEB vol. 3, p. 5].

21 **"My Wife Popsy Wopsy":** Edison technical notes and drawings, Feb. 14, 1872 [TAED NE1678059].

21 **"I like it first-rate":** *Philadelphia Weekly Times*, April 27, 1878, p. 3 [TAED SM029054a].

21 **She slept with a revolver:** "The Wizard of Menlo Park, by His Daughter Marion Edison Oser" [n.d.], p. 6 [TAED X018A5Z].

22 **"I wouldn't give a penny":** Dyer and Martin (1910:767–68).

22 **"That's wrong!":** Tate (1938:126).

22 **"I am the Professor of Physics":** Barker to Edison, Nov. 3, 1874 [TAED NM003132; also in TAEB vol. 2, p. 328].

23 **"The man who is certain he is right":** Bence Jones (1870:310).

23 **"I think I must have tried":** Dyer and Martin (1910:101).

24 **"Mr. Edison . . . promises to become":** *New York Herald*, Dec. 2, 1875, p. 4 [TAED PA074; also in TAEB vol. 2, p. 668].

24 **"Mr. Edison has named":** *New-York Tribune*, Nov. 30, 1875, p. 2.

24 **"a sort of first cousin to electricity":** *Daily Graphic* (New York), Dec. 2, 1875, p. 3 (vol. 9, p. 243).

24 **"may prove of great value":** *Boston Daily Globe*, Dec. 1, 1875, p. 4.

25 **"I do not hesitate to pronounce":** *Scientific American*, Feb. 19, 1876, p. 116.

25 **"simply gratuitous":** *Scientific American Supplement*, Jan. 29, 1876, p. 78.

25 **"Edison About to Astonish the World Again":** *Telegrapher*, July 29, 1876, pp. 184–85.

CHAPTER 3: NEMESIS

27 **University of Michigan's Detroit Observatory:** For a history of the observatory, see Whitesell (1998).

28 **"He knew the stars":** Frieze et al. (1882:31).

28 **"Two or three or four times a year":** *Indianapolis News*, Aug. 2, 1878, p. 2.

28 **he telegraphed his discovery to Joseph Henry:** Watson to Henry, Sept. 28, 1876 [SIA Accession 11-032, Reel M161, Control No. 26884].

28 **wired the news to Europe:** *Nature* 14 (363), Oct. 12, 1876, p. 540.

29 **"so small that a good walker":** *Scientific American Supplement*, June 23, 1877, p. 1224.

29 **"vermin of the skies":** Peebles (2000:28).

30 **"[W]e have great hope that yours is a new one":** Peters to Watson, Nov. 9, 1857 [JCWP Box 1].

30 **"I wish they would stop awhile":** Watson to Peters, Nov. 16, 1857 [CHFPP Box 0000.123.2.1].

30 **Eurydice:** When naming their planets, Peters and Watson used the English spelling of names from classical mythology, but the International Astronomical Union today recognizes the German spelling for asteroids discovered in that time period. Hence Peters's Eurydice is officially Eurydike, Watson's Clymene is Klymene, etc. See Kirkwood (1888), Schmadel (2003).

31 **"I can surrender my claims":** *Detroit Free Press*, Oct. 18, 1865, p. 1.

31 **"Dr. PETERS gracefully yields":** *Hamilton Campus*, Oct. 31, 1868, p. 2.

31 **"I congratulate you"**: Peters to Watson, Sept. 23, 1868 [JCWP Box 1].

31 **"The score stands"**: *Michigan University Magazine*, Nov. 1868, p. 73.

31 **"Of his personality it may be said"**: Newcomb (1903:373).

31 **"in defiance of rule"**: Holden (1896:30).

32 **"great planet shooting match"**: *Detroit Post*, April 21, 1876, p. 2.

32 **"If these fellows go on"**: *Atlanta Daily Sun*, Sept. 27, 1871, p. 3.

32 **"The picking up of asteroids"**: *Daily Graphic* (New York), Aug. 20, 1877, p. 2 (vol. 14, p. 334).

32 **"Prof. P. is beating Prof. Watson"**: *Michigan Argus* (Ann Arbor), May 30, 1873, p. 2.

32 **"selfish and unscrupulous in advancing his own interests"**: Charles Young to George Comstock, Aug. 22, 1887 [GADRUWM Series 7/4/2 Box 4].

32 **"The Hon. James C. Watson one of the greatest"**: 1855 notebook, p. 62 [JCWP Box 2]. For more on Watson's college notebook, see Rufus (1938).

CHAPTER 4: "PETTICOAT PARLIAMENT"

34 **"Large Republican Gains"**: *North American* (Philadelphia), Oct. 4, 1876, p. 1.

34 **"I think that the war"**: *Press* (Philadelphia), Oct. 4, 1876, p. 8.

34 **"The principal event of yesterday"**: *Philadelphia Inquirer*, Oct. 4, 1876, p. 2.

35 **elegant hall of pillars and frescoes**: *Philadelphia Inquirer*, March 25, 1876, p. 2.

35 **portrait of Queen Victoria**: The portrait, painted by Thomas Sully, now hangs at the Metropolitan Museum of Art in New York City. For more, see Barratt (235).

35 **Woman's Congress, an annual symposium**: The gathering, sometimes called the *Women's* Congress, was organized by the Association for the Advancement of Woman, alternately called the Association for the Advancement of *Women*. For clarity's sake, I consistently refer to the annual meeting using the singular possessive: Woman's Congress.

35 **including Julia Ward Howe**: Howe wrote about the meeting in her diary entries of Oct. 4–5, 1876 [HFP Item 1107, vol. 11]. See also J. W. Howe (1884:458–59), J. W. Howe (1900:387), J. W. Howe (1906:288), and AAW (1893:7).

35 **She stepped to a desk**: Details of the first day of the Woman's Congress come from accounts published on Oct. 5, 1876, in many Philadelphia newspapers: *Evening Telegraph*, p. 8; *Philadelphia Inquirer*, p. 4; *Press*, p. 2; *Public Ledger*, p. 1; *Public Record*, p. 1; *Times*, p. 1. Also *Daily Graphic* (New York), Oct. 6, 1876, p. 2 (vol. 11, p. 664); *Woman's Journal*, Oct. 14, 1876. The original program for the 1876 congress is in MMM Reel 7, Item 59.

35 **"When we inquire"**: In recounting Mitchell's speech, I have relied upon the transcription in the meeting's official proceedings [AAW (1877:9–11)]. The speech also appeared, with minor differences in wording, in *Woman's Journal*, Oct. 14, 1876, p. 332; and *Daily Rocky Mountain News* (Denver), Oct. 29, 1876, p. 1.

36 **"Astronomy enters into the price"**: Thompson and Rodgers (1878:2).

36 **"As it is 'Venus'"**: Davis to Mitchell, Jan. 7, 1851 [USNO-LOC Box 15].

37 **"The lightning that he caught"**: Maria Mitchell Notes, 1880, p. 29 [MMM Reel 3, Item 24].

38 **"O'Brian's Belt"**: Recorded by Nathaniel Hawthorne in his personal notebook; see Woodson (1980:51). Mitchell, who had not met Hawthorne before arriving in Europe, traveled with his family from Paris to Rome. "[S]he seems to be

a simple, strong, healthy-humored woman," the famous author wrote of the celebrated astronomer [Woodson (1980:17)]. She wrote of him: "I found Mr. Hawthorne very taciturn" [Kendall (1896:90)].

38 **"[I]t was evident he did not":** MMM Reel 8, in "Maria Mitchell Letters," after Item 69a. Slightly edited version in Kendall (1896:143).

38 **"[T]he Father kindly informed me":** MMM Reel 2, Item 17. Transcribed in Kendall (1896:158). The astronomer-priest was Angelo Secchi; see Mazzotti (2010).

38 **an injustice she fought:** Booker (2007:322–26).

38 **"Beautiful Venus, pride of the morning":** "Miss Mitchell Sleeps in a Bed," handwritten poem by Elizabeth Owen Abbot [MMP Folder 6.8]. Typed version with slightly different wording appears in MMM Reel 4, Item 34.

39 **"When arrested development":** E. H. Clarke (1873:92–93).

39 **"death from over-work":** E. H. Clarke (1873:103).

39 **"Vassar victims":** *Vassar Miscellany*, Jan. 1874, p. 126; *Medical News*, Oct. 14, 1882, p. 437.

39 **barrage of rebuttals:** Comfort and Comfort (1874), Duffey (1874), J. W. Howe (1874), Jacobi (1877).

39 **"If we know the number of young girls":** *Woman's Journal*, Oct. 23, 1875, p. 341.

40 **"full and positive equality":** AAW (1877:26).

40 **"crush and dishonor":** AAW (1877:77).

40 **"tatterdom of flimsy, frayable":** AAW (1877:98).

40 **"Here were assembled":** *Chicago Times*, Oct. 18, 1875, p. 4. See also *Woman's Journal*, Nov. 13, 1875, p. 368.

40 **as a "petticoat parliament":** *Syracuse Daily Courier*, Oct. 20, 1875, p. 4. See also *World* (New York), Oct. 17, 1873, p. 6.

40 **"Women are needed in scientific work":** AAW (1877:9).

41 **"Does anyone suppose":** AAW (1877:11).

41 **the audience had difficulty hearing:** In her diary, Mitchell complained of the poor acoustics. See entry of Nov. 15, 1876 [MMM Reel 6, Item 50, p. 84; transcribed in Albers (2001:243)].

41 **under the title: "The Need of Women in Science":** *Daily Rocky Mountain News* (Denver), Oct. 29, 1876, p. 1.

CHAPTER 5: POLITICS AND MOONSHINE

45 **The Babylonians, Greeks, Mayans, and Chinese:** Stephenson (1997:58–61), Westfall and Sheehan (2015:108–12), Aaboe et al. (1991), Littmann et al. (2008:33–37).

46 **It was not until the eighteenth century:** G. Armitage (1997).

46 **"for therby the Situation and dimensions":** Pasachoff (1999:2.19). See also Halley (1715), Gingerich (1981). For more on Halley, see A. Armitage (1966), Cook (1998).

47 **it gave him a headache:** Brewster (1855:157–58).

48 **"a big, lusty, joyous man":** F. C. Howe (1925:30).

48 **"piercing eyes, a look full of strength":** James Bryce, in Wead et al. (1910:136).

48 **"like a glacier":** Merrick (1910:687).

48 **obsessed with the motion of the moon:** Newcomb (1903:202–11), Norberg (1978).

48 **"Miss M. is only a female astronomer after all"**: Newcomb to William P. G. Bartlett, May 20, 1862 [SNP Box 4].

48 **"the orbit of any one planet"**: C. Wilson (1989:253).

48 **"To this work I was especially attracted"**: Newcomb (1903:63).

49 **"a rather dilapidated old dwelling-house"**: Newcomb (1903:214). The office soon moved to more respectable quarters at Washington's Corcoran Building.

49 **"I wish to issue some information"**: Newcomb to Hill, Jan. 16, 1878 [USNO-LOC Box 16].

49 **"Professor Newcomb says Mr. Hill"**: Journal of Mabel Loomis, entry for May 3, 1878, p. 224 [MLTP Series III, Box 45, Folder 47, Microfilm Reel 7].

49 **"to compute the central line"**: Newcomb to Hill, Jan. 23, 1878 [USNO-LOC Box 16].

49 **"During its progress, the dark shadow"**: NAO (1878:3).

50 **lukewarm review from Maria Mitchell**: *Christian Union*, April 17, 1878, p. 328.

50 **"Total eclipses of the sun afford"**: Newcomb (1878*b*:29).

50 **"[A]s the last ray of sunlight vanishes"**: Newcomb (1878*b*:252).

50 **"Besides this 'corona'"**: Newcomb (1878*b*:252).

51 **"We thus have the seeming paradox"**: Newcomb (1878*a*:849).

51 **"[W]e will never by any means"**: Translated from Comte (1835:8). See also M. Pickering (1993), Hearnshaw (2010).

51 **"As the geologist with his hammer"**: *New-York Tribune*, Jan. 15, 1873, p. 4.

53 **British astronomer Warren De La Rue**: De La Rue (1862).

53 **Lockyer named it helium**: Jules Janssen has often been described as the co-discoverer of helium, but recent books argue that Lockyer deserves sole credit. See Launay (2012:45–46), Nath (2013:5).

53 **"that my body had become a kind of projectile"**: Tyndall (1873:433).

53 **struck rocks off Sicily**: There are many accounts of the shipwreck. See W. G. Adams (1871:155), T. M. Lockyer et al. (1928:58), Roscoe (1906:158–62).

53 **escaped by balloon**: Launay (2012:49–51).

54 **"as a Friend of Science"**: Baxter (1914:402).

54 **Williams blamed the latter**: Williams (1785:102). See also Gingerich (1981), Rothschild (2009).

55 **"The country was practically under water"**: Newcomb (1903:92). Newcomb's diary from 1860 includes details of the journey [SNP Box 1; see also Kennedy and Hanson (1996)]. Newcomb creatively recounted the expedition in an unpublished manuscript titled "Up the Saskatchewan" [SNP Box 108]. Scudder (1886) also offers an amusing narrative of the trip, albeit with the names disguised.

55 **collected insect, fish, and fossil specimens**: Unsigned article by Newcomb in *The Nor'Wester* (Red River Settlement, Canada), Aug. 14, 1860, p. 3.

56 **like an ear of corn**: USNO (1885:124).

56 **a slew of findings**: Sands (1870).

56 **"[T]he American Government and men of science"**: *Nature* 1 (1), Nov. 4, 1869, p. 15.

56 **"[O]ne of our scientific men"**: Newcomb (1903:114).

56 **"It is doubtful whether we shall have any"**: Newcomb to Lindsay, Jan. 2, 1878 [SNP Box 4].

57 **"[I]t is still uncertain whether Congress"**: Newcomb to Lockyer, Nov. 18, 1877 [SNP Box 4; also at SJNL].

57 **occasional guest at the Hayes White House**: Journal of Mabel Loomis, entry

for Feb. 10, 1878, pp. 173–74 [MLTP Series III, Box 45, Folder 47, Microfilm Reel 7], mentions one White House reception that Newcomb attended. Newcomb's daughter Anita wrote school essays about two family visits to the Hayes White House [ANMP Box 8].

57 **soon-to-be President James A. Garfield:** Letters from Garfield to Newcomb can be found in SNP Box 22. Newcomb's letters to Garfield are in the James A. Garfield Papers, also at the Library of Congress. In 1881, when the assassin Charles Guiteau shot Garfield, Newcomb assisted Alexander Graham Bell in trying to locate where in the president's body the bullet had lodged, and Newcomb devised a scheme to cool the White House, both futile attempts to save Garfield's life. See Newcomb (1903:353–63); "List of Journeys," pp. 22–28 [SNP Box 123]; Temkin and Koudelka (1950); *New York Herald*, July 15, 1881, p. 2; various items in AGBFP Box 221.

57 **proposed sending seven government parties:** Rodgers to Congressman John D. C. Atkins, March 19, 1878 [USNO-NA Entry 4, vol. 3, pp. 568–71; also USHR Records of Legislative Proceedings, Committee on Appropriations: Navy (HR45A-F3.10), Folder 8].

58 **"The sun is the source of all light":** "The desirability of Observing Eclipses of the Sun," March 15, 1878, an attachment to letter from Rodgers to Congressman John D. C. Atkins [USNO-NA Entry 4, vol. 3, pp. 566–68; also USHR Records of Legislative Proceedings, Committee on Appropriations: Navy (HR45A-F3.10), Folder 10].

58 **"This seems to be another instance":** *Denver Daily Tribune*, April 20, 1878, p. 2.

59 **"The lame excuse":** *New-York Tribune*, April 8, 1878, p. 4. Also see article on p. 1.

59 **"[W]e should feel in a very awkward condition":** *Congressional Record* (1878:2857).

59 **"[Senator Blaine] proposes to send a commission":** *Congressional Record* (1878:2858).

CHAPTER 6: THE WIZARD IN WASHINGTON

60 **all was green along the National Mall:** *Independent Statesman* (Concord, N.H.), April 25, 1878, p. 1.

60 **"the increase and diffusion of knowledge":** Bruce (1987:187).

60 **infested the building with fleas:** Bruce (1987:299).

60 **"noon repast":** *Syracuse Daily Journal*, April 23, 1878, p. 2.

61 **"about the sleepiest and slowest institution":** Newcomb to Henry Draper, April 20, 1877 [HMAPDP]. For more on the early years of the National Academy of Sciences, see Kevles (2013).

61 **"The name of the animal":** *Omaha Daily Bee*, July 9, 1878, p. 4.

62 **her portrait stared down from the wall:** *Cincinnati Commercial*, April 22, 1878, p. 2.

62 **three-story brick row house:** "Our home on Eleventh St.," by Anita Newcomb McGee [SNP Box 124].

62 **"I shall be only too glad to do anything":** Barker to Newcomb, April 7, 1878 [SNP Box 57].

62 **blind man's buff:** Barker to Newcomb, April 28, 1878 [SNP Box 57]. This was a

favorite game in the Newcomb household and was often played by Barker when in town. See "Blind-mans-buff," a school essay by Newcomb's daughter Anita, Jan. 14, 1878 [ANMP Box 8]; Journal of Mabel Loomis, entry for May 11, 1878, p. 230 [MLTP Series III, Box 45, Folder 47, Microfilm Reel 7]; Barker to Mrs. Newcomb, April 13, 1898 [SNP Box 15].

62 **liked to braid his whiskers:** Barker to Newcomb, March 25, 1907 [SNP Box 15].

62 **ten minutes past four:** Edison's presentation had been scheduled for 4:00 P.M. but began ten minutes late [NAS, *N.A.S. Minutes 1863–1882*, pp. 532–34].

63 **"stood out at all angles in defiance of comb rule":** *Evening Star* (Washington, D.C.), April 19, 1878, p. 1 [TAED MBSB10537b]. Many details of the presentation of Edison's phonograph and telephone come from this same article, as well as from the following: *Washington Post and Union*, April 19, 1878, p. 1 [TAED SM029004a] and p. 4 [TAED MBSB10535X]; *New-York Tribune*, April 20, 1878, p. 6 [TAED MBSB10543X]; *Daily Critic* (Washington, D.C.), April 20, 1878, p. 4; *Cincinnati Commercial*, April 22, 1878, p. 2; *Evening News* (Detroit), April 23, 1878, p. 2; *Albany Evening Journal*, April 23, 1878, p. 1; *Syracuse Daily Journal*, April 23, 1878, p. 2.

63 **"The speaking phonograph has the honor":** *Washington Post and Union*, April 19, 1878, p. 4 [TAED MBSB10535X].

63 **"I declare," remarked a member of the audience:** *Evening Star* (Washington, D.C.), April 19, 1878, p. 1 [TAED MBSB10537b].

63 **"The Napoleon of Science":** *Sun* (New York), March 10, 1878, p. 6 [TAED SB031032b].

63 **"The Jersey Columbus":** *Daily Graphic* (New York), April 2, 1878, p. 1 (vol. 16, p. 221) [TAED MBSB10472; also in TAEB vol. 4, p. 213].

63 **"The Wizard of Menlo Park":** *Daily Graphic* (New York), April 10, 1878, p. 5 (vol. 16, p. 281) [TAED SB031090a].

63 **Menlo Park as "Edisonia":** *Daily Graphic* (New York), April 2, 1878, p. 1 (vol. 16, p. 221) [TAED MBSB10472; also in TAEB vol. 4, p. 213].

63 **"The Mania has broken out":** George Bliss to Edison, April 13, 1878 [TAED D7805ZAL; also in TAEB vol. 4, pp. 229–30].

64 **"The idea of a talking machine":** *Sun* (New York), April 29, 1878, p. 3 [TAED MBSB10561].

64 **"Speech has become, as it were, immortal":** *Scientific American*, Nov. 17, 1877, p. 304 [TAED SM030022a].

64 **"Let me, like all the rest of the world":** *Washington Post and Union*, April 19, 1878, p. 1 [TAED SM029004a].

65 **demonstrated his phonograph to congressmen:** *Evening Star* (Washington, D.C.), April 19, 1878, p. 1 [TAED MBSB10537a]; *Cincinnati Commercial*, April 20, 1878, p. 5; *Philadelphia Inquirer*, April 20, 1878, p. 1 [TAED SM029014a]; *Sun* (Baltimore), April 20, 1878, p. 1; *Raleigh News*, April 21, 1878, p. 1; *Albany Evening Journal*, April 23, 1878, p. 1; *Independent Statesman* (Concord, N.H.), April 25, 1878, p. 1; *Milwaukee Daily News*, April 25, 1878, p. 1; *Daily Rocky Mountain News* (Denver), April 26, 1878, p. 2.

65 **late night drop-in at the White House:** *Chicago Daily Tribune*, May 4, 1878, p. 12; Jehl (1937:159–60); Dyer and Martin (1910:210); TAEB vol. 4, p. 859 and p. 863.

65 **aversion to crowds:** Dyer and Martin (1910:108–9).

65 **"a man of deeds, not words":** *Evening Star* (Washington, D.C.), April 19, 1878, p. 1 [TAED MBSB10537b].

66 **"Halloo, halloo"**: *Record* (Philadelphia), April 20, 1878, p. 1 [TAED SM029007a].

66 **this word as the standard telephone greeting**: Kennelly (1933:293), Koenigsberg (1987).

66 **"You do well, Mr. Edison"**: *Washington Post and Union*, April 19, 1878, p. 4 [TAED MBSB10535X].

67 **"not very long ago . . . became notorious"**: *New-York Times*, March 25, 1878, p. 4 [TAED MBSB10456X].

67 **"I think that science is the greatest interest"**: *Cincinnati Enquirer*, April 20, 1878, p. 5.

67 **"I suppose this odor is used"**: Edison to Darwin, Dec. 7, 1877 [TAED Z002AA; also in TAEB vol. 3, p. 657].

68 **"I used to hold my hands up"**: Goode (1897:204).

68 **"If you could make something"**: Langley to Edison, Dec. 3, 1877 [TAED D7702ZDL; also in TAEB vol. 3, p. 651].

69 **"Have you made any recent improvements"**: *Evening Star* (Washington, D.C.), April 19, 1878, p. 1 [TAED MBSB10537b].

CHAPTER 7: SIC TRANSIT

70 **"There has been an unusual interest"**: *Chronicle* (Univ. of Michigan), April 20, 1878, p. 195.

70 **"[T]he Naval Observatory desires, if possible"**: John Rodgers to Watson, Jan. 5, 1878 [JCWP Box 1].

71 **"Next Monday the astronomers of Europe and this country"**: *New-York Tribune*, May 4, 1878, p. 4.

71 **encyclopedias and textbooks**: *The National Encyclopaedia* (1867:655), Campbell (1909:496).

71 **"The Fourth of July must return"**: *New-York Times*, Sept. 26, 1876, p. 4.

72 **"Professor Peters of Hamilton college"**: *Connecticut Courant* (Hartford), Oct. 12, 1876, p. 2.

72 **long expressed doubts about Vulcan's existence**: Newcomb (1860).

72 **"Prof. Watson has long been a believer"**: *Chronicle* (Univ. of Michigan), Oct. 12, 1878, p. 3.

72 **"Mercury has, or has been thought to have"**: *Inter Ocean* (Chicago), May 4, 1878, p. 4.

73 **"Even members of the [French] Academy"**: Newcomb (1903:328). See also Lequeux (2013).

73 *au bout de sa plume*: Académie des Sciences (1846:660).

74 **"[D]uring the last ten or fifteen years"**: *Utica Weekly Herald and Gazette and Courier*, May 20, 1873, p. 2.

74 **Copernicus was said never to have seen it**: Lynn (1892).

74 **"whether the result of LE VERRIER"**: USNO (1878b:3).

75 **"the exact instant when an unseen spherical body"**: M. Mitchell (1869:558).

75 **magnified the heavens by a factor of 400**: Many details of Watson's observations derive from "Report of Observations of the Transit of Mercury, May 6, 1878 at Ann Arbor, Michigan," by James C. Watson [USNO-NA Entry 18, Box

1; draft report in JCWP Box 2]. Further details from *Detroit Free Press*, May 8, 1878, p. 1; *Ann Arbor Register*, May 8, 1878, p. 3.

75 **Mitchell used the transit as a teaching opportunity:** *Poughkeepsie Daily Eagle*, May 7, 1878, p. 3. Mitchell's observing notes are in MMP Folder 11.2. Mitchell described her transit observations in her annual report to Vassar's president, John Raymond ["Astronomical Department, May 21, 1878," VCA Folder 1.37]. She also mentioned them in her diary entry of May 6, 1878; see Albers (2001:248) [original in MMM Reel 6, Item 50, p. 95]. Mitchell's glass negatives of the transit are in MMP Series IX.

75 **Peters pointed. "There, professor," he said:** *Utica Weekly Herald and Gazette and Courier*, May 7, 1878, p. 5.

76 **borrowed a telescope to view the event:** A. K. Eaton to Edison, May 4, 1878 [TAED D7802ZKE1]; Norman C. Miller to Edison, May 6, 1878 [TAED D7802ZKF].

76 **"Sick transit," he said, "glorious Monday":** *New-York Tribune*, May 7, 1878, p. 5. See also *New York Herald*, May 7, 1878, p. 4; *New-York Times*, May 7, 1878, p. 5; USNO (1879:25–36). The observing notebook of Edward S. Holden, who joined Draper and Barker for the transit, is in USNO-NA Entry 18, Box 1.

77 **"[T]he planet appeared as a sharply defined":** USNO (1879:58).

77 **"The fact that the planet made":** *Detroit Free Press*, May 8, 1878, p. 1.

77 **"Dr. Peters reiterates his disbelief":** *Albany Evening Journal*, May 9, 1878, p. 2; *Milwaukee Daily Sentinel*, May 14, 1878, p. 5; *Independent Statesman* (Concord, N.H.), May 16, 1878, p. 262.

77 **"As the truth of Leverrier's discovery":** USNO (1878a:8).

78 **"Dear Sir: I have just heard":** Rodgers to multiple recipients, April 29, 1878 [USNO-NA Entry 4, vol. 4, p. 29]. The copy of the letter received by Newcomb is clearly in Edison's electric pen [USNO-NA Entry 24, Box 6]. A follow-up letter from Rodgers dated May 29, 1878, has been explicitly noted as copied via the electric pen [USNO-NA Entry 4, vol. 4, p. 61].

78 **"accepts" or "regrets":** Replies are in USNO-LOC Box 8.

78 **"It is a great temptation":** Peters to Edward S. Holden, June 8, 1878 [LOR Box 36, Folder 1].

78 **the Shoshone-Bannocks:** Heaton (2005).

78 **scrapped its planned expeditions to western Montana:** Rodgers to Rear Admiral Daniel Ammen, June 4, 1878 [USNO-NA Entry 2, vol. 3, p. 485]; USNO (1880:145).

78 **"I accept your polite invitation":** Watson to Rodgers, July 13, 1878 [USNO-LOC Box 8]. Watson was late to receive an invitation because the USNO had originally assumed he would be out of the country at the time of the eclipse; see Rodgers to Watson, July 11, 1878 [USNO-NA Entry 4, vol. 4, p. 108].

CHAPTER 8: "GOOD WOMAN THAT SHE ARE"

80 **she had met the prior week:** The date was April 10, 1878; see *Inter Ocean* (Chicago), March 30, 1878, p. 9. Mitchell wrote in her diary about the meeting and seeing Joseph Henry [MMM Reel 6, Item 50, pp. 90–93; partially transcribed in Albers (2001:247)].

80 **"Our meeting will be for a few hours":** Mitchell to Henry, Oct. 25, 1877 [SIA Record Unit 26, vol. 168, pp. 94–95; partially transcribed in Albers (2001:247)].

81 **"not . . . publish an account"**: Henry to Mitchell, Oct. 31, 1877 [SIA Record Unit 33, vol. 58, p. 239].

81 **Maria Mitchell inquired if her students might participate**: Mitchell sent her initial inquiry to Joseph Henry, who forwarded it to C. H. Davis. See Mitchell to Henry, March 3, 1874 [USNO-NA Entry 18, Box 49].

81 **"[I]t would be absolutely out of the question"**: Davis to Mitchell, March 17, 1874 [USNO-NA Entry 18, Box 54].

82 **10,82 to 15,82 pounds**: Telegram from Edward S. Holden to Quartermaster General's Office, June 20, 1878 [USNO-LOC Box 8].

82 **specially guarded freight car**: USNO (1880:323). One of the USNO assistants who accompanied the car westward, Henry Smith Pritchett (later president of MIT), recalled the journey in an unpublished manuscript, "The Chronicles of Henry Smith" [HSPP Box 1].

82 **"fitted up in a style"**: *New-York Times*, Sept. 14, 1875, p. 5. This article refers to the same train car (Pennsylvania Railroad Postal Car No. 10) used by the USNO in 1878. See *Philadelphia Inquirer*, Sept. 14, 1875, p. 3; also "Bills of Lading for Insts. sent West" [USNO-LOC Box 13].

82 **extended courtesies to foreign astronomers**: *Nature* 17 (440), April 4, 1878, p. 453; *Monthly Notices of the Royal Astronomical Society* 38 (6), April 12, 1878, pp. 335–37.

82 **"Will you please inform me"**: Mitchell to Rodgers, June 3, 1878 [USNO-LOC Box 8].

82 **appeal from an amateur astronomer in Fort Dodge**: Hess to Rodgers, June 3, 1878 [USNO-LOC Box 8].

82 **inspired by Simon Newcomb's** *Popular Astronomy*: Hess to Newcomb, April 3, 1907 [SNP Box 26].

82 **Rodgers obliged him**: Rodgers to Ticket Agents of Union Pacific and other Railroads, June 6, 1878 [USNO-NA Entry 4, vol. 4, pp. 71–72].

82 **"It is very certain that if any one"**: Rodgers to Mitchell, June 7, 1878 [USNO-NA Entry 4, vol. 4, pp. 72–73].

82 **"It is time that women worked in earnest"**: Mitchell to Emily Talbot, Jan. 19, 1878 [BMCL Letters and Documents Collection].

83 **"The college is handsomely decorated"**: *Sunday Courier* (Poughkeepsie), June 23, 1878, p. 3.

83 **Professor Mitchell's annual "dome party"**: Details of the 1878 dome party come from *Vassar Miscellany*, July 15, 1878, pp. 14–15. Broader recollections of the event's traditions can be found in the following: *Vassar Miscellany*, July 1876, p. 781; *Vassar Miscellany*, July 1877, pp. 250–51; *Vassar Miscellany*, Jan. 1909, pp. 199–200; Abbott (1889); Kendall (1896:191–92); Babbitt (1912:22–26); *In Memoriam: Maria Mitchell*, by Mary W. Whitney [MMP Folder 14.5]; "Maria Mitchell & Vassar College – Club Paper," by Mary A. Mineah [MMP Folder 14.11]; "Maria Mitchell's Dome Parties," by Martha (Hillard) MacLeish [MMP Folder 6.7]; "Maria Mitchell," by Rebecca W. Hawes, pp. 24–26 [MMM Reel 4, Item 34].

84 **"I shall always believe that strawberries"**: *Woman's Journal*, Sept. 2, 1882, p. 277.

84 **"Here's to our Jessie"**: MMP Folder 6.10.

84 **"We are singing for the glo-ry"**: Vassar College (1881:94–95). The lyrics have been published many other times, with slight variations.

85   **"[T]he English language was ransacked"**: *Woman's Journal*, July 6, 1889, p. 213.

85   **"I cannot expect to make astronomers"**: MMM Reel 3, Item 28. Transcribed in Kendall (1896:184).

86   **final work for her seniors**: Blatch and Lutz (1940:39).

86   **science was best taught outside the classroom**: AAW (1877:10).

86   **a similar expedition a decade earlier**: *Burlington Daily Hawk-Eye*, Aug. 5, 1869, p. 4; *Burlington Daily Hawk-Eye*, Aug. 6, 1869, p. 4; *Burlington Daily Hawk-Eye*, Aug. 10, 1869, p. 2; M. Mitchell (1869); USNO (1885:55–58).

86   **Mitchell declined to comment**: *Burlington Daily Hawk-Eye*, Aug. 8, 1869, p. 4. Mitchell was induced, however, to write a magazine article about her eclipse expedition; see Albers (2001:179–81), M. Mitchell (1869).

86   **"What a magnificent opportunity"**: Blatch and Lutz (1940:39).

87   **"the biggest disappointment of my life"**: Blatch and Lutz (1940:39).

87   **"Social Prejudices against Woman's Entering"**: *Vassar Miscellany*, July 1877, p. 256.

87   **Culbertson helped organize the astronomy class**: Culbertson to Miss Swan, June 16, 1907 [MMP Folder 14.3].

87   **Cora Harrison, spoke on behalf of the class**: Harrison to President Raymond, June 25, 1877 [MMP Folder 14.3].

87   **possessed her own telescope**: *Daily Rocky Mountain News* (Denver), July 24, 1878, p. 4. Harrison later bequeathed her telescope to Vassar for the use of students; see *Vassar Miscellany*, Jan. 1889, p. 133.

87   **"Oh C. W. M."**: "Dome Party, June 12, 1873" [MMM Reel 4, Item 34].

87   **"[I] can truly say that I know"**: Mitchell to Abbot, Feb. 1874 [MMP Folder 1.13].

88   **"I hope yet to be as simply, comfortably"**: Kendall to Henrietta Wolcott, June 7, 1875 [NEWCR Reel 14].

88   **"[W]e were a party of six"**: MMM Reel 4, Item 33. Transcribed in Kendall (1896:225).

CHAPTER 9: SHOW BUSINESS

89   **"As a wonderful inventor, he is himself red-hot"**: *Daily Graphic* (New York), June 4, 1878, p. 2 (vol. 16, p. 666).

89   **"[I]t would require a Fahrenheit thermometer"**: *New-York Tribune*, May 29, 1878, p. 5.

90   **The inventor smiled and obligingly bowed**: *New-York Times*, June 4, 1878, p. 5 [TAED MBSB10687X].

90   **"Heat causes the strip of hard rubber to expand"**: *New York Herald*, June 4, 1878, p. 10 [TAED MBSB10644X].

91   **more fully displayed its capabilities**: *New York Herald*, June 22, 1878, p. 3 [TAED MBSB10694].

91   **"In this way it is not improbable"**: *New York Herald*, June 4, 1878, p. 10 [TAED MBSB10644X].

91   **"No name has yet been given"**: *New-York Tribune*, June 4, 1878, p. 5.

91   *Micro-thermo-meter. Micro-thermo-scope*: TAED D7835V (image 3).

92   **Edison suggested installing it in lighthouses**: *Daily Graphic* (New York), April 2, 1878, p. 3 (vol. 16, p. 223).

92   *telephonoscope* **also projected the voice**: *Boston Daily Globe*, May 24, 1878, p.

2 [TAED SM029088b]; *Boston Post*, May 24, 1878, p. 3 [TAED SM029092a]. Some other reports called it a *telescopophone*. Edison later took to calling the device his *megaphone*; *Sun* (New York), June 8, 1878, p. 3 [TAED SM029144].

92 **"I wonder if you couldn't talk a hole"**: *New York Herald*, April 24, 1878, p. 7 [TAED SM029018b]. Somewhat different accounts of how and where Edison was inspired to invent the phonomotor are in *Sunday Courier* (Poughkeepsie), May 5, 1878, p. 2; *Boston Evening Transcript*, May 23, 1878, p. 4 [TAED SM029086b]; *Boston Morning Journal*, May 25, 1878, p. 1 [TAED SM029096b].

92 **"It is expected to become a favorite method"**: *Chicago Daily Tribune*, May 4, 1878, p. 12.

93 **"one of the most accomplished, ingenious"**: Hodgson et al. (1885:207). See more at Cranston (1993).

93 **one of the few believers**: Blavatsky (1877:126–27).

93 **1877's "blue glass craze"**: Pleasonton (1876); *New York Herald*, March 8, 1877, p. 5 [TAED SM005002a]; *Scientific American*, April 7, 1877, p. 208.

93 **Solid Muldoon**: *Colorado Weekly Chieftain* (Pueblo), Oct. 4, 1877, p. 2; *New-York Tribune*, Jan. 24, 1878, pp. 1–2. For more on hoaxes of the era, see Pettit (2006), Tribble (2009).

94 **"[T]he midgets or dwarfs draw $1000⁰⁰ houses"**: George H. Bliss to Edison, June 15, 1878 [TAED D7829ZCF].

94 **Barnum expressed interest**: Uriah H. Painter to Edison, May 1, 1878 [TAED D7802ZJM].

94 **"It degrades the machine"**: Redpath to George H. Smardon, June 11, 1878 [TAED X154B4BH]. For background on Redpath, see McKivigan (2008).

94 **the frontier drama**: R. A. Hall (2001), Kasson (2000), Sagala (2008). For an exploration of the life and career of J. B. "Texas Jack" Omohundro, see Logan (1954).

94 **"There is much absurdity in the action"**: *Boston Evening Journal*, June 4, 1878, p. 1.

95 **"I received a shot in my stirrup"**: *New York Herald*, Sept. 1, 1877, p. 10. See also *Sun* (New York), Oct. 22, 1877, p. 1.

95 **told the papers a different tale**: *Sioux City Weekly Journal*, Oct. 11, 1877, p. 3.

95 **"Instead of being termed 'Texas Jack, the well known scout'"**: *Helena Independent*, Sept. 12, 1877, p. 3.

95 **"ONE OF THE CURIOUS THINGS about the visit"**: *Boston Herald Supplement*, June 1, 1878, p. 2 [TAED SM029137b]. See also *Daily Evening Traveller* (Boston), May 23, 1878, p. 2 [TAED SM029106a]; *Boston Evening Transcript*, May 23, 1878, p. 4 [TAED SM029086b]; *Boston Daily Advertiser*, May 24, 1878, p. 4 [TAED SM029092b]; *Boston Morning Journal*, May 24, 1878, p. 4; *Boston Daily Globe*, May 24, 1878, p. 2 [TAED SM029088b]; *Boston Post*, May 24, 1878, p. 3 [TAED SM029092a]; *Boston Morning Journal*, May 25, 1878, p. 1 [TAED SM029096b]; *Boston Daily Globe*, May 28, 1878, p. 2 [TAED SM029117a].

96 **"Well, is it perfected yet?"**: *Record* (Philadelphia), June 5, 1878, p. 4 [TAED MBSB10648].

96 **"I expect to go in the beginning of July"**: Langley to Edison, June 22, 1878 [TAED D7802ZPR]. See also *Daily Post* (Pittsburgh), June 27, 1878, p. 4.

96 **one from Chicago**: This was Elias Colbert. See George H. Bliss to Edison, July 3, 1878 [TAED D7802ZRG].

96 **duo from Princeton:** The pair consisted of Charles A. Young and Cyrus F. Brackett. See *New-York Tribune*, June 8, 1878, p. 10 [TAED MBSB10665b]. See also Young to Edison, June 10, 1878 [TAED D7835A].

97 **Edison accused the man, David Hughes:** *Sun* (New York), June 9, 1878, p. 7 [TAED MBSB10670]. For more on Hughes, see I. Hughes and D. E. Evans (2011).

97 **"Mr. Edison says that the bores":** *Sun* (New York), May 18, 1878, p. 2 [TAED MBSB10595X].

98 **"I have prayed for an earthquake":** *Daily Graphic* (New York), April 11, 1879, p. 3 (vol. 19, p. 295) [TAED MBSB21163].

98 **"I am pretty badly used up":** *Daily Graphic* (New York), June 13, 1878, p. 2 (vol. 16, p. 730) [TAED MBSB10674a].

98 **"Edison is laid up for repairs":** *Daily Constitution* (Atlanta), July 5, 1878, p. 4.

98 **the White Mountains, the Great Lakes, the Atlantic shore:** William K. Applebaugh to Edison, June 14, 1878 [TAED D7802ZOZ]; George H. Bliss to Edison, June 25, 1878 [TAED D7829ZCG]; Uriah H. Painter to Edison, June 17, 1878 [TAED D7802ZPC].

98 **In April, he had approached Simon Newcomb:** Barker to Newcomb, April 7, 1878 [SNP Box 57].

98 **"I have heard nothing from you":** Newcomb to Barker, June 7, 1878 [SNP Box 4].

99 **"This morning I was up in time to see Venus":** Henry Draper to Daniel Draper, Aug. 3, 1862 [JWDFP Box 23].

99 **"I will try and call at your place":** Edison to Draper, Aug. 8, 1877 [TAED X120BAL; also in TAEB vol. 3, p. 489].

99 **"I am under so many obligations":** Morton to Edison, Feb. 5, 1878 [TAED D7802ZAH].

99 **"Have seen Professor Barker":** Edison to Draper [n.d.] [TAED D7802ZUC].

99 **"The latest marvel from Menlo Park":** *New York Herald*, June 22, 1878, p. 3 [TAED MBSB10694].

100 **Simon Newcomb advised that the odds of clear skies:** Newcomb to Draper, June 4, 1878 [USNO-LOC Box 17]. Within a few weeks, Draper had shifted his destination to Rawlins; see Edward S. Holden to Samuel P. Langley, June 21, 1878 [SPLP].

100 **"Edison is coming West":** *San Francisco Chronicle*, July 10, 1878, p. 3.

100 **"EDISON, the inventor, of phonograph fame":** *Daily Rocky Mountain News* (Denver), June 27, 1878, p. 4.

100 **"Now that Mr. Edison has joined":** *Press* (Philadelphia), July 16, 1878, p. 4. See similar comment in *Denver Daily Tribune*, July 21, 1878, p. 4, col. 5.

100 **ex-telegrapher friend of Edison's:** *Brooklyn Daily Eagle*, Feb. 20, 1905, Picture Section, p. 3; P. Jones (1940:161).

101 **"If I should fall back":** Dickson and Dickson (1894:234). The cast was made for *The Phrenological Journal and Life Illustrated*; see H. S. Drayton to Edison, July 1, 1878 [TAED D7802ZQZ].

101 **shipped nearly a ton of scientific equipment:** Barker (1878:104).

101 **Pacific Express train:** Barker to Edison, July 12, [1878] [TAED D7802ZSG; also in TAEB vol. 4, p. 393].

101 **"If the sun's corona has any heat":** *Daily Graphic* (New York), July 19, 1878, p. 1 (vol. 17, p. 121).

CHAPTER 10: AMONG THE TRIBES OF UNCIVILIZATION

105   **"The eye has no joy":** Bowles (1869:54).

105   **"I arrived here about 12.30 P.M.":** A. N. Skinner to Rear Admiral John Rodgers, July 14, 1878 [USNO-NA Entry 7, Box 45].

106   **tin cans and the skull of a cow or buffalo:** Visible in E. L. Trouvelot's illustration of Creston eclipse camp [USNO-NA Entry 18, Box 8]. See image in chapter 12.

106   **A small graveyard lay nearby:** Marked on "Ground Plan of Eclipse Parties at Creston, Wyoming," by E. L. Trouvelot [USNO-NA Entry 18, Box 8].

106   **makeshift observatory:** Skinner to William Harkness, July 9, 1878 [USNO-NA Entry 18, Box 8]; USNO (1880:30).

106   **ordered logistical help:** USNO (1880:xiii).

106   **four-mule wagon:** Special Order No. 66, July 13, 1878 [USACC Entry 8, vol. 4].

106   **fresh supplies of ice and hay:** Major Thomas Tipton Thornburgh to William Harkness, July 23, 1878 [USNO-NA Entry 18, Box 8]; Lt. George N. Chase to William Harkness, July 22, 1878 [USNO-NA Entry 18, Box 8].

106   **"The ride to Harrisburg":** Diary of James E. Keeler, July 19, 1878 [JEKP Box 1]. After graduating from Johns Hopkins, Keeler rose to prominence as a respected astrophysicist. See Osterbrock (1984).

106   **"simply dreadful—I've never seen":** David Peck Todd to Mabel Loomis, July 10, 1878 [MLTP Series I, Box 2, Folder 76].

107   **"There is a pretty rough crowd":** Diary of James E. Keeler, July 21, 1878 [JEKP Box 1].

107   **"endeavored to get up a flirtation":** *Chicago Daily Tribune*, July 27, 1878, p. 5. For more on Texas Jack's hunting party, see *Omaha Daily Bee*, July 22, 1878, p. 4; *Burlington Daily Hawk-Eye*, July 23, 1878, p. 8; *Forest and Stream*, Oct. 24, 1878, p. 241.

107   **"[J]ust *think* where I am":** David Peck Todd to Mabel Loomis, July 11, 1878 [MLTP Series I, Box 2, Folder 76].

108   **Special wheels—made of compressed paper:** Helena E. Wright (1992). For more on Pullman's opulent cars, see Husband (1917); Leyendecker (1992); Welsh, Howes, and Holland (2010).

108   **an exclusive "hotel car":** *Wheeling Daily Register*, July 17, 1878, p. 2.

108   **"We sip our oyster-soup":** *Frank Leslie's Illustrated Newspaper*, Aug. 25, 1877, p. 422.

109   **celebrities, including the ubiquitous Texas Jack:** *Inter Ocean* (Chicago), Aug. 9, 1876, p. 8.

109   **cavernous, frescoed lobby:** *Frank Leslie's Illustrated Newspaper*, July 21, 1877, p. 344.

109   **felling not only people but also the horses:** *Chicago Daily Tribune*, July 16, 1878, p. 1; *Chicago Times*, July 17, 1878, p. 1; *Chicago Daily Telegraph*, July 17, 1878, p. 1.

109   **a scrum of reporters:** *Chicago Times*, July 16, 1878, p. 6.

109   **"Have you ever been in Chicago before?":** *Inter Ocean* (Chicago), July 16, 1878, p. 8.

109   **"How many patents have you now?":** *Chicago Daily Tribune*, July 16, 1878, p. 5.

109   **"The tasimeter is a heat-measurer":** *Chicago Daily Tribune*, July 16, 1878, p. 5.

110 **"Well, then, begin your fusillade"**: *Omaha Daily Bee*, July 17, 1878, p. 4.

110 **"to ride on the Locomotive"**: J. J. Dickey, July 17, 1878 [TAED D7802ZSW; also in TAEB vol. 4, p. 397].

110 **"without dust or anything else"**: Dyer and Martin (1910:231).

111 **"Saw Pikes Peak. 160 miles away"**: Henry Morton 1878 daybook, entry for July 18, 1878 [PSIT Box 3].

111 **"a big bridge for a small brook"**: "Reminiscences," by Hezekiah Bissell, p. 43 [WSA Microfilm H-36].

112 **cried "Hands up!"**: *Laramie Daily Sentinel*, May 30, 1878, p. 4; *Council Bluffs Daily Nonpareil*, May 31, 1878, p. 4; *Laramie Daily Sentinel*, June 11, 1878, p. 2.

112 **"Four Masked Men Clean Out"**: *Philadelphia Inquirer*, May 31, 1878, p. 1.

112 **"The company will pay $1,112"**: *Laramie Daily Sentinel*, June 4, 1878, p. 3.

112 **A posse soon captured the men**: *Cheyenne Weekly Leader*, June 6, 1878, p. 1.

112 **final destination around midnight**: *New-York Times*, July 26, 1878, p. 5; Barker (1878:104). Edison in later years recalled, incorrectly, that he had arrived in Rawlins at "about 4 P.M." [Dyer and Martin (1910:227)].

112 **Superintendent Dickinson met the scientific party**: J. J. Dickey to Edison [July 18, 1878] [TAED D7802ZSW1].

113 **"I will be at Rawlins on Sunday night"**: Fox [to Draper?], [n.d.] [TAED D7802ZUD]. Fox likely sent this telegram on July 19, 1878, from Omaha, where he had stopped for a day or two. See *Omaha Daily Bee*, July 20, 1878, p. 4.

113 **"good fare and commodious rooms"**: *Carbon County News* (Rawlins), July 13, 1878, p. 1 (advertisement).

113 **thirteen in all**: *Daily Independent* (Laramie), Oct. 5, 1874, p. 3.

113 **jamborees, charity balls, and concerts**: *Laramie Daily Sentinel*, April 17, 1875, p. 3; *Laramie Daily Sun*, Sept. 22, 1875, p. 4; *Laramie Daily Sentinel*, March 9, 1877, p. 4; *Carbon County News* (Rawlins), March 2, 1878, p. 1; Vivian (1879:323).

113 **stabbed a young man**: *Laramie Daily Sentinel*, Aug. 19, 1876, p. 4.

113 **"the boss pistol-shot of the West"**: As recalled by Edison in Dyer and Martin (1910:227). That Texas Jack made this statement rings true; see *Burlington Daily Hawk-Eye*, July 23, 1878, p. 8; *Times* (London), Aug. 27, 1878, p. 6.

113 **"The shot awakened all the people"**: Edison in Dyer and Martin (1910:228). Edison made a similar statement in *The New York Herald*, May 7, 1916, First Section–Part 4, p. 5.

113 **Texas Jack was reportedly deep in debt**: *Daily Illinois State Register* (Springfield), Oct. 10, 1877, p. 2; *Brooklyn Daily Eagle*, March 17, 1878, p. 4.

113 **his marriage was suffering**: Thorp (1957:90).

113 **"Both Fox and I were so nervous"**: Dyer and Martin (1910:228). Edison seemed to recall that the encounter occurred on his first night in Rawlins, but that is not possible. He arrived with the Draper party the night of July 18. Fox reached Rawlins the night of July 21 [*Laramie Daily Sentinel*, July 22, 1878, p. 4]. Texas Jack did not get to town until the night of July 23 [*Laramie Daily Sentinel*, July 24, 1878, p. 4].

114 **"Rawlins Red"**: "Lillian Heath, M. D.," by Neal Miller, p. 13 [WSA Microfilm H-5]; *Salt Lake Weekly Tribune*, June 22, 1878, p. 1.

114 **"Rawlins presents to the curious eye"**: *Frank Leslie's Illustrated Newspaper*, Nov. 3, 1877, p. 138.

114 **lumberyard that moonlighted in coffins**: *Carbon County Journal* (Rawlins), Feb. 28, 1880, p. 1 (advertisement).

114 **one-story stone structure with three small cells:** *Laramie Daily Sun*, Sept. 27, 1875, p. 4.

114 **a horse thief and three of the four men:** The horse thief was Richard Duff; the train robbers were W. A. Gibson, William Henry, and Dick Hill. See *Cheyenne Weekly Leader*, June 6, 1878, p. 1; "Jail Register, Carbon County, Wyoming Territory" [CCM]; Frye (1990:55–56).

114 **The fourth had turned state's witness:** This was John Thomas. See *Laramie Daily Sentinel*, June 4, 1878, p. 3; Frye (1990:56).

114 **"He looked like a 'bad man'":** Dyer and Martin (1910:228).

114 **afternoon excursion to Brown's Canyon:** Henry Morton 1878 daybook, entry for July 22, 1878 [PSIT Box 3].

114 **from a merchant in Laramie:** Receipt, "Ed. Dickinson Bought of Louis Miller," July 20, 1878 [TAED D7807AA (image 11, bottom)].

114 **offered his house and yard:** *New-York Tribune*, July 26, 1878, p. 8; *Nature* 18 (461), Aug. 29, 1878, p. 464; Galbraith (1922); Craig (1931:33).

114 **William Daley, a carpenter who owned the Rawlins lumberyard:** *Daily Independent* (Laramie), Nov. 24, 1874, p. 3; *Laramie Daily Sun*, May 17, 1875, p. 4; *Rawlins Republican*, Sept. 28, 1922, p. 8; Bartlett (1918:144–49).

115 **Lillian Heath, an inquisitive twelve-year-old:** "Lillian Heath, M. D.," by Neal Miller, p. 11 [WSA Microfilm H-5].

115 **"to see the red man on his native heath":** *Greeley Tribune*, Aug. 7, 1878, p. 2.

115 **After the eclipse, he went camping:** *Laramie Daily Sentinel*, Aug. 12, 1878, p. 4; *Laramie Daily Sentinel*, Aug. 13, 1878, p. 4.

115 **"The-chief-who-shoots-the-stars":** *Harper's Weekly*, Oct. 25, 1879, p. 844.

115 **"You could throw a crumb of bread":** *World* (New York), Aug. 27, 1878, p. 1 [TAED SB032075a].

115 **barrel stave driven through his mouth:** Dawson and Skiff ([1879]1980:56). For more on the background, unfolding, and aftermath of the Meeker Massacre and the Battle of Milk Creek, in which Major Thornburgh was killed, see Meeker ([1879]1976), M. D. Moody (1953), Emmitt ([1954]2000), Sprague (1957), M. E. Miller (1997), Decker (2004), Silbernagel (2011).

117 **in front of Fred Wolf's saloon:** Entry for "Parrot George" in "Jail Register, Carbon County, Wyoming Territory" [CCM].

117 **"some of the best people of the town":** *Carbon County Journal* (Rawlins), March 26, 1881, p. 4. Details of the lynching are from this article and from the coroner's inquest of March 23, 1881 [WSA Parrott, George: RG1030, Carbon Co. Clerk of District Court, Series 07.01, Coroner's Inquest].

117 **precociously assisted with the autopsy:** *Rawlins Daily Times*, May 12, 1950, pp. 1 and 16; *Rawlins Daily Times*, May 13, 1950, p. 1; *Denver Post*, Aug. 28, 1955, *Empire Magazine*, p. 7; *Montana: The Magazine of Western History*, vol. 28, No. 1 (Winter 1978), p. 84; L. Brown (1995).

118 **fashioned into a pair of Oxfords:** *Rawlins Republican*, Sept. 27, 1925, p. 1; Ridenour (2008). The shoes are on display at CCM.

CHAPTER 11: QUEEN CITY

119 **three hospitals, eight banks:** Corbett, Hoye and Co. (1878); "Thayer's Map of Denver, Colorado," 1879 [DPLWHC CG4314 .D4 1879 .T49].

120 **"Let the echo go out":** *Denver Daily Times*, May 20, 1872, p. 2.

120 **the owner of Denver's successful Inter-Ocean Hotel:** F. Hall (1895:440–41); *Denver Post*, Aug. 3, 1969, *Empire Magazine*, pp. 20–25.

120 **Chinatown-destroying riot:** Wortman (1965).

121 **"Why, Coloradoans are the most disappointed":** Strahorn (1878:67).

121 **"Sir, Colorado can beat":** *Times* (London), Dec. 27, 1878, p. 8.

121 **"Tourists are coming into Denver":** *Sunshine Courier*, July 20, 1878, p. 2. See also *Daily Rocky Mountain News* (Denver), July 14, 1878, p. 4; *Denver Daily Tribune*, July 27, 1878, p. 4; *Chicago Times*, July 30, 1878, p. 1.

121 **bunked on a billiard table:** *Denver Daily Times*, July 23, 1878, p. 4.

121 **judges, U.S. senators:** *Denver Daily Tribune*, July 21, 1878, p. 4.

121 **petty thieves from New York:** *Laramie Daily Sentinel*, July 26, 1878, p. 4.

121 **"thick as blackberries":** Vickers (1880:237).

121 **"A cannon shot fired in any direction":** *Evening Star* (Washington, D.C.), July 23, 1878, p. 2.

121 **arrived from stately Princeton:** Details of the party and its activities are in Young (1878a); Ranyard (1881); C. G. Rockwood's solar eclipse record book, 1878 [ASDR Box 7, Folder 8]; *New-York Tribune*, July 27, 1878, p. 2; *Daily Rocky Mountain News*, July 28, 1878, p. 4; *Denver Daily Times*, July 29, 1878, p. 4; *Denver Daily Tribune*, July 30, 1878, p. 4; *The Princetonian*, Oct. 10, 1878, pp. 83–84.

122 **Known affectionately to his students as "Twinkle":** Frost (1910:105).

122 **"We were all charmed with Prof. Young":** Sara Glazier Bates to Mrs. Hinchman, June 11, 1904 [MMM Reel 4, Item 34].

123 **Arthur Cowper Ranyard—stayed:** Before departing England, Ranyard had received—and accepted—an invitation to join Charles Young's camp at Denver. See Ranyard to Henry Draper, June 11, 1878 [HMAPDP]; Ranyard to Young, July 3, 1878 [CAYP Box 7].

123 **high school astronomy class:** C. G. Rockwood's solar eclipse record book, 1878, entry for "Wednesday 17th" [ASDR Box 7, Folder 8].

123 **Chief Colorow was, with reason, bitter:** Silbernagel (2011:208–10).

123 **"The [Princeton] students were especially interested":** *Rocky Mountain News* (Denver), June 2, 1918, p. 12.

124 **"The eclipse will be total here":** *Denver Daily Times*, July 17, 1878, p. 2.

124 **"At the time of the eclipse the star Procyon":** *Daily Rocky Mountain News*, July 27, 1878, p. 4.

124 **"[I] will endeavor to show my *patriotism*":** James C. Pratt to Newcomb, July 1, 1878 [USNO-NA Entry 24, Box 6].

125 **invited volunteers to attend a class:** Chicago Astronomical Society (1878); *Denver Daily Times*, July 26, 1878, p. 4; *Daily Rocky Mountain News* (Denver), July 26, 1878, p. 4; *Daily Rocky Mountain News* (Denver), July 27, 1878, p. 4; *Chicago Daily Tribune*, Aug. 2, 1878, p. 3; *Chicago Times*, Aug. 3, 1878, p. 2. Professions of class members found in Corbett, Hoye and Co. (1878).

125 **"Many persons went down to their graves":** *Daily Rocky Mountain News* (Denver), July 19, 1878, p. 2.

125 **"Wouldn't it be an excellent idea":** *Denver Daily Times*, July 26, 1878, p. 1.

126 **"His name is Edison, the great inventor":** *Denver Daily Tribune*, July 25, 1878, p. 4.

126 **Rumors circulated that he would join the Princeton party:** *Daily Rocky Mountain News* (Denver), July 10, 1878, p. 4.

126 **Denver Press Club planned a special reception:** *Daily Rocky Mountain News* (Denver), July 9, 1878, p. 4.

126 **"After all the expectation":** *Denver Daily Tribune*, July 21, 1878, p. 4.

126 **"Mr. Edison is doubtless":** *Daily Rocky Mountain News* (Denver), June 30, 1878, p. 2.

126 **"hour after hour and day after day":** MMM Reel 4, Item 33. Transcribed in Kendall (1896:224).

127 **"One peculiarity in travelling from East to West":** Kendall (1896:57).

127 **"Thirty-three women and children":** Dall (1881:19).

127 **"at the expense of her future usefulness":** Van de Warker (1872:204). For more on women and nineteenth-century rail travel, see Richter (2005).

127 **"I am thirteen years old":** George M. Flick to Mitchell, Sept. 23, 1874 [MMM Reel 8, Item 68].

128 **fought in the courts and on the land:** Athearn (1962).

128 **"We learned that there was a war":** MMM Reel 4, Item 33. Transcribed in Kendall (1896:226).

128 **The Denver & Rio Grande was refusing to carry:** *Daily Rocky Mountain News* (Denver), July 6, 1878, p. 2.

128 **"[W]ar, no matter where or when":** MMM Reel 4, Item 33. Transcribed in Kendall (1896:226).

128 **Cora Harrison '76 and Emma Culbertson '77—who had arrived earlier:** *Denver Daily Tribune*, July 24, 1878, p. 4.

128 **"This party adds peculiar interest":** *Sun* (New York), Aug. 4, 1878, p. 2.

129 **Colorado had considered a radical proposal:** J. G. Brown (1898), B. B. Jensen (1973), Mead (2004).

129 **the indefatigable Susan B. Anthony:** Gordon (1997:318–25).

129 **"How absurd and revolting":** *Denver Daily Tribune*, Jan. 23, 1877, p. 4. For more on Machebeuf, see Howlett ([1908]1954).

129 **"We must 'keep pegging away'":** *Woman's Journal*, Feb. 9, 1878, p. 45.

130 **the daughter of a New York abolitionist:** Vickers (1880:306).

130 **"fear the physician":** Norris (1915:62).

130 **"I grind my teeth in despair over it":** Avery to Caroline Wells Healey Dall, Nov. 17, 1873 [CWHDP Reel 7 (Box 5, Folder 22)]. Also see J. W. Howe (1874:191–95).

130 **they pressed Vassar's board:** Booker (2007:322–26), Albers (2001:200–2).

130 **ten thousand dollars a year:** *Daily Graphic* (New York), Oct. 25, 1877, p. 6 (vol. 14, p. 802).

130 **called upon to treat Josephine Meeker:** Schurz et al. (1880:44).

130 **"Have you a bit of land":** MMM Reel 4, Item 33. Transcribed in Kendall (1896:224).

130 **gracious two-story home surrounded by a rose garden:** Sweet (1894:637), Dall (1881:25).

131 **A frequent visitor was Helen Hunt Jackson:** In diary entries from 1876 through 1878, Jackson recorded several visits to Avery's house [HHJP Part 1, Ms 0020, Box 5].

131 **"[I]t is just lovely":** Gordon (1997:325). See also Harper (1899:493), and Anthony's diary entries for Oct. 3–23, 1877 [SBAP Reel 2].

131 **"Homes of Single Women":** SBAP Reel 7.

131 **Avery now hosted the Vassar eclipse party:** *Daily Rocky Mountain News*

(Denver), July 28, 1878, p. 4; *Daily Rocky Mountain News* (Denver), Oct. 12, 1908, p. 4.

132 **the luggage containing the other tube and the lenses:** MMM Reel 4, Item 33. Transcribed in Kendall (1896:225–26).

CHAPTER 12: NATURE'S EDITOR

133 **"So many circumstances . . . have to be noted":** Smyth (1853:503).

134 **America's foremost manufacturer of optical instruments:** Warner and Ariail (1995).

134 **"[T]heir artistic execution is excellent":** Walker (1880:497). For more on Trouvelot and his drawings, see Trouvelot (1882), Corbin (2007).

134 **As he later told Alvan Clark:** Kirkland (1906:9).

134 **one of the most devastating insect invasions:** Spear (2005).

134 **At Creston, life for Trouvelot, Clark:** Details of camp life are in *The New York Herald*, July 30, 1878, p. 3 [OBSERVATIONS OF THE NAVAL OBSERVATORY PARTY IN WYOMING . . .]. This article was evidently written by William Harkness; see similar language in his eclipse report in USNO (1880:29–73).

137 **"a barren wilderness of sand":** Newcomb to his wife, July 21 [1878] [SNP Box 12]. Newcomb further described the camp, its activities, and setting in his eclipse report in USNO (1880:99–116) and in his 1878 eclipse expedition notebook [SNP Box 2].

137 **surveyors plotting the transcontinental railroad parted ways:** Crofutt (1878:102), Shearer (1882–83:94).

137 **The rough building, sixteen feet long:** *New-York Tribune*, July 26, 1878, p. 8. See also Barker (1878); *Nature* 18 (461), Aug. 29, 1878, pp. 462–64.

137 **"The preparations for observing":** *New York Herald*, July 29, 1878, p. 5 [TAED MBSB10799X].

138 **the Draper party took time out for an excursion:** The Draper party's visit to Separation occurred on Wednesday, July 24, as noted by Newcomb in his 1878 eclipse expedition notebook [SNP Box 2] and by Morton in his 1878 daybook [PSIT Box 3], although Morton miswrote, "Went to Division [rather than Separation] to see Prof. Newcomb." Many years later, Edison vividly recalled the excursion [Dyer and Martin (1910:229–30)] but claimed, incorrectly, that it occurred two days before the eclipse; i.e. Saturday, July 27. J. J. Clarke (1929) also colorfully recounted the visit but miswrote that it occurred *after* the eclipse. Other descriptions on which this scene is based are in *The World* (New York), Aug. 27, 1878, p. 1 [TAED SB032075a]; *Sun* (New York), Aug. 29, 1878, p. 3 [TAED SB032060a]; *Frank Leslie's Budget*, March 1888, p. 75; *Rawlins Republican*, Sept. 28, 1922, p. 8; *Arkansas Gazette* (Little Rock), Nov. 1, 1931, Features Section, p. 1. The day after the gathering at Separation, Newcomb wrote in a postcard to his wife: "Yesterday we were honored with a visit from the Rawlins party comprising Draper, Barker, Edison & Morton. Showed them how to eat lunch with a chisel and drink out of a tin dipper & wooden pail. Dined them in our best style." [SNP Box 13]

139 **"Well, that's one on me":** F. A. Jones (1908:303).

139 **"I have a sick boy at home":** Lockyer to Newcomb, June 18 [1878] [SNP Box 30].

139 **"Since my last note to you":** Lockyer to Newcomb, July 3 [1878] [SNP Box 30].

139 **"The military furnish my party":** Newcomb to Lockyer, July 15, 1878 [SNP Box 4; also at SJNL].

140 **He had recently chided Henry Draper:** J. N. Lockyer (1878c:86). For more on Draper's supposed discovery, see Plotkin (1977).

140 **"Sky colours impossible":** *Nature* 18 (445), May 9, 1878, p. 31.

140 **"And Lockyer, and Lockyer":** Cortie (1921:241). Other variants of this poem were also in circulation. *Times* (London), May 9, 1878, p. 5: "Of the solar corona/ He says, 'I'm the owner,'/ And sneers at the moon as far rockier." Armstrong (1928:871): "There was a young astronomer called Lockyer,/ Who each year grew cockier and cockier,/ Till he thought he was owner of the solar corona,/ Did this young astronomer Lockyer." Bain (1940:142): "Than Mr. Lockyer/ Nothing can be cockier,/ He talks of the corona/ As if he were the owner."

140 **"Lockyer sometimes forgets":** Macfarlane (1916:101). For more on Lockyer, see Meadows (2008).

141 **"for whose scientific attainments":** *New-York Tribune*, July 19, 1878, p. 2.

141 **Lockyer sent a dispatch to Newcomb:** Noted in Newcomb's 1878 eclipse expedition notebook on July 23 [SNP Box 2].

141 **Lockyer crossed into Wyoming in a Pullman car:** Ryan (1879:26).

141 **"[S]ome of us will meet you at Rawlins":** Newcomb to Lockyer, July 24 [1878] [SNP Box 57].

141 **"There was quite a party":** "Diary of Mrs. Simon Newcomb on first trip to Europe," p. 4 [SNP Box 2].

141 **checked in at the Railroad Hotel:** For a list of names on the hotel's guest register as of July 25, 1878, see "SOL IN THE SHADE. Fred Hess Gives an Account of His Expedition," unidentified clipping presumed to be from *Webster County Gazette and Messenger* (Fort Dodge), Aug. 16, 1878 [USNO-NA Entry 18, Box 8].

141 **borrowed from Michigan State Normal School:** Watson in USNO (1880:117). The telescope remains today in the possession of Eastern Michigan University.

142 **"[She] is in every thought devoted":** *Cincinnati Commercial*, Aug. 16, 1878, Supplement, p. 1.

142 **"[H]is treatment of his wife":** Young to George Comstock, Aug. 22, 1887 [GADRUWM Series 7/4/2 Box 4].

142 **the Taj Mahal by moonlight:** Watson's "Journal of voyage &c," Jan. 19 to March 27, 1875, entry for March 25, p. 69 [JCWP Box 1].

142 **the Great Pyramid of Giza:** Watson's diary of 1875 to Egypt and Arabia, entry for April 18, pp. 13–15 [JCWP Box 1].

142 **climbing Mt. Vesuvius:** Annette Watson to "My dear little sister," Aug. 5, 1875 [JCWP Box 1].

142 **riding gondolas:** Annette Watson to "Dear Sister Pussie," Aug. 25, 1875 [JCWP Box 1].

142 **"I do enjoy traveling & sightseeing":** Annette Watson to "Dear Parents," May 10, 1875 [JCWP Box 1].

142 **"J. C. is so nervous & almost sick":** Annette Watson to "Dear Parents," Feb. 11, 1875 [JCWP Box 1].

142 **"He has just told me":** Annette Watson, undated fragment of letter [JCWP Box 1].

142 **Daniel Hector Talbot, a prosperous land broker:** Stephens (1944:20–26); *Morning Oregonian* (Portland), Sept. 3, 1878, p. 3; *Iowa City Daily Press*, Feb. 5, 1912, p. 8; Talbot to Edison, Oct. 14, 1878 [TAED D7802ZZEX].

142   **W. Fraser Rae, a British journalist:** *New-York Tribune*, Aug. 13, 1878, p. 4 and p. 5 [TAED MBSB10819b].

143   **"At a little after midnight arrived":** R. C. Lehmann's travel diary, entry for July 26, 1878 [LFP Box 132, Folder 2]. Years later, Lehmann's experience in Wyoming influenced his literary work. In a fictionalized account of English university life, he described a Cambridge don who was noted for "having discovered a new planet during a recent eclipse of the sun" [Lehmann (1897:57)].

143   **J. B. Silvis was a former saloonkeeper:** Swackhamer (1994).

143   **"[He] meanders up and down":** *Papillion Times* (Nebraska), reprinted in *Anthony's Photographic Bulletin*, Feb. 1875, p. 53.

144   **THE ECLIPSE STATION AT RAWLINS, WYOMING TERRITORY:** Identification of individuals based on *Sioux City Journal*, Feb. 8, 1914, p. 7, which labels the fourth man from the left "Edison's friend, Bloomfield," yet there is no record in the Edison Papers of a Bloomfield among the inventor's associates. The same article identifies the fifth man from the left as "Prof. S. Hess, Fort Dodge, Ia." This is presumably F. Hess, who wrote about J. B. Silvis taking the group photograph in "Sol in the Shade. Fred Hess Gives an Account of His Expedition," unidentified clipping presumed to be from *Webster County Gazette and Messenger* (Fort Dodge), Aug. 16, 1878 [USNO-NA Entry 18, Box 8]. Galbraith (1922) provides different names for some individuals in the photograph—and his identifications have been copied by several later writers—but Galbraith is a notably unreliable source. (See epilogue.) The photograph was taken by J. B. Silvis on Saturday, July 27, 1878, according to *The Standard* (London), Aug. 16, 1878, p. 2.

144   **"He was the worst dressed man":** *Salt Lake Daily Tribune*, Aug. 14, 1878, p. 1.

145   **invited the people of Rawlins to look through their telescopes:** "Sol in the Shade. Fred Hess Gives an Account of His Expedition," unidentified clipping presumed to be from *Webster County Gazette and Messenger* (Fort Dodge), Aug. 16, 1878 [USNO-NA Entry 18, Box 8]; *Rawlins Republican-Bulletin, Union Pacific Old Timers Edition*, May 2, 1939, p. 42.

145   **could spot the moons of Jupiter:** *Nature* 18 (460), Aug. 22, 1878, p. 431.

145   **"[The scientists] never tired":** *Laramie Daily Sentinel*, July 30, 1878, p. 4.

145   **"[Draper's] great reputation rests":** R. C. Lehmann's travel diary, entry for July 28, 1878 [LFP Box 132, Folder 2].

145   **they all reminisced, boasting:** *New York Herald*, July 29, 1878, p. 5 [TAED MBSB10799X].

145   **Lockyer recalled his last eclipse, in India:** J. N. Lockyer (1874:332–55).

146   **in Peking for the transit of Venus:** Watson's report on the transit of Venus is in Newcomb (1881:103–13).

146   **the sun represented the emperor:** Lu and Li (2013).

146   **Emperor Tongzhi fell sick, diagnosed with smallpox:** Chang (2013:97–110).

146   **only by good luck and strategy:** Upon his immediate return from China, Watson downplayed any personal risk [*Michigan Argus* (Ann Arbor), Oct. 29, 1875, p. 3]. In later retellings by Watson and others, the story was embellished, presumably to heighten the drama [*Sun* (New York), April 22, 1881, p. 1].

146   **"This being the first planet":** Watson in Newcomb (1881:108). See also Watson to Joseph Henry, May 28, 1877 [SIA Record Unit 26, vol. 166, pp. 528–29]. Charles Young recalled Juewa's discovery in a letter to George Comstock, Aug. 22, 1887 [GADRUWM Series 7/4/2 Box 4].

147 **He hired a Rawlins carpenter:** Watson (1878a:230); USNO (1880:117–18); *Nature* 18 (461), Aug. 29, 1878, p. 462.

147 **"The approach of any person":** *New York Herald*, July 29, 1878, p. 5 [TAED MBSB10799X]. Other details of Edison's tasimeter setup are in Edison (1879), Barker (1879).

148 **"It's strange," Edison mumbled:** Fox (1879:451–52). In this account, written months after the test, Fox misstated that it took place on the "night previous to the eclipse." Fox's earlier article in *The New York Herald*, July 29, 1878, p. 5 [TAED MBSB10799X], as well as Edison (1879) and other sources, reported that it occurred on Saturday, July 27.

148 **"One of the many points of interest":** *Nature* 18 (459), Aug. 15, 1878, pp. 401–2.

CHAPTER 13: OLD PROBABILITIES

150 **"Rain?—oh, no, it doesn't 'rain' in Colorado this year":** *Colorado Miner* (Georgetown), July 27, 1878, p. 3.

150 **a lifelong fear of lightning:** Kendall (1896:16); *Poughkeepsie Daily Eagle*, Aug. 12, 1893, p. 6.

150 **almost swamped the tents:** C. G. Rockwood's solar eclipse record book, 1878, entry for "Monday 22nd" [ASDR Box 7, Folder 8].

150 **"which is precisely what every one who has faith":** *Denver Daily Times*, July 29, 1878, p. 1.

151 **"hovered over our fair city":** Weeks (1878:260).

151 **"Probable northeast to southwest winds":** Fenno (1878:137).

151 **"so that neither Moon nor Stars":** Labaree (1961:464).

152 **information was posted at the Smithsonian:** Maury (1871:401); *Monthly Weather Review*, June 1898, pp. 263–64.

152 **sought a new peacetime role:** C. Abbe (1895:234). See more in Whitnah (1961), Raines (1996).

153 **"giving the presence, the course":** Barnes et al. (1870:4).

153 **"[W]hen mother had taken me on her lap":** "What I remember about my friends and myself," by Cleveland Abbe, p. 6 [CAP-LOC Box 29]. For more on Abbe's early life and later career, see Humphreys (1919), T. Abbe (1955).

154 **postgraduate fellow in Russia:** Reingold (1963), Reingold (1964). For more on Pulkovo Observatory, where Abbe worked, see Werrett (2010).

154 **a search that proved an exasperating struggle:** Hetherington (1976).

154 **"It was physical-, mental-":** Abbe to Newcomb, March 17, 1867 [SNP Box 14].

154 **"I am determined to show the autocrat":** Mitchel (1887:53).

154 **marveled at the 1806 total solar eclipse:** C. F. Adams (1874:442–43).

154 **"lighthouses of the skies":** Bemis (1956:503).

154 **one of the Greek columns was tipping over:** *Cincinnati Commercial*, May 4, 1869, p. 4; Historical and Philosophical Society of Ohio (1944:44).

155 **the area's worsening air pollution:** C. Abbe (1869).

155 **beginning in September 1869, he offered the service:** C. Abbe (1871).

155 **"Old Probabilities," or "Old Probs":** C. Abbe (1916:207).

155 **It had once been home to James Monroe:** M. B. Morris (1918).

155 **"toys which excite the envy":** *Chicago Daily Tribune*, June 14, 1873, p. 5.

155 **messages were condensed using a cipher:** USASC (1877:351, 349, 342).

156 **an average of one hour and forty minutes:** USASC (1877:137).

156 **stuck in a military hierarchy:** T. Abbe (1955:185), Greely (1916:703).

156 **"anything relating in the remotest degree":** Abbe to Barker (draft), Jan. 27, 1876 [CAP-LOC Box 7]. The final letter is in GFBP Box 1, Folder 19.

157 **long walks and croquet:** T. Abbe (1955:106).

157 **rigid and domineering:** *Daily Critic* (Washington, D.C.), Feb. 7, 1880, p. 1 [AJMP Reel 3]; *Popular Science Monthly*, Jan. 1881, p. 409.

157 **it should conduct basic research:** Reingold (1964:146).

157 **Myer evinced little interest in theoretical studies:** Humphreys (1919:480), Whitnah (1961:38–42).

157 **usurped his nickname:** Evidence of Abbe's resentment that his boss had assumed his nickname can be found in Abbe's archived files [CAP-LOC Box 31]. On Aug. 31, 1880—one week after General Myer's death—*The National Republican* (Washington, D.C.) reported: "The popular impression that 'Old Probabilities' died at Buffalo a few days since is an erroneous one." The article went on to explain that the genuine "Old Probabilities" had come to the Signal Service from Cincinnati "in the person of Professor Cleveland Abbe." Abbe clipped and saved this news item, and he may well have been the reporter's source of information.

157 **"Somehow I always feel better":** Cleveland Abbe to Fanny Abbe, Aug. 25, 1874 [CAP-LOC Box 7].

157 **"I shall be greatly obliged":** Rodgers to Myer, Feb. 4, 1878 [USNO-NA Entry 4, vol. 3, p. 541].

158 **The Signal Service issued these figures:** C. Abbe (1881:14–17). A list of recipients can be found in WB NC-3, Entry 12, vol. 2, p. 338.

158 **"The totality had passed away":** *Cincinnati Daily Gazette*, Aug. 14, 1869, p. 2. For more on Abbe's eclipse expedition, see *Cincinnati Commercial*, Aug. 14, 1869, p. 5; *Cincinnati Daily Enquirer*, Aug. 14, 1869, p. 8; C. Abbe (1872). Abbe's 1869 diary, in which he recorded details of the journey, is in CAP-LOC Box 1.

158 **"[O]ur party was one of the very few":** Myer in Sands (1870:193–96).

159 **highest-altitude weather station on the planet:** USASC (1889:vii). See also Smith (1993).

159 **hoped to go to Pikes Peak himself:** As late as two weeks before the eclipse, Myer was still not sure if he would go. On July 16, 1878, he wrote a letter to Norman Lockyer (for the Englishman to receive when he arrived in New York) extending an invitation to observe together in Colorado: "I think of going myself and I beg to offer you the hospitality of the Peak. . . . Your going would make my doing so almost a certainty." [SJNL]

159 **large holes in the bedspreads and carpeting:** Dall (1881:53).

159 **"This morning the sky is everywhere":** Cleveland Abbe to Fanny Abbe, July 21, 1878 [CAP-LOC Box 8]. Transcribed in T. Abbe (1955:218).

160 **"I am very glad to hear from you":** Langley to Abbe, Nov. 5, 1868 [CAP-LOC Box 4].

160 **"It's first-rate;—your going":** Langley to Abbe, June 17, 1878 [CAP-LOC Box 8].

160 **"[General Myer] takes pleasure in offering":** H. H. C. Dunwoody to Langley, June 13, 1878 [WB NC-3, Entry 6, vol. 13, p. 471]. Also in C. Abbe (1881:26), with minor differences in wording.

160 **"there was no room for [either of] us":** From Langley's official eclipse report in USNO (1880:203–10). He also described his activities on Pikes Peak in Langley (1891:50–59) and in *Pittsburgh Commercial Gazette*, Aug. 13, 1878, p. 1. Abbe's

official report is in C. Abbe (1881:42–56). These sources provided most details for the remainder of this chapter.

161 **passed waterfalls and crags and snowfields:** Descriptions of the trail to the summit are in *The Chicago Times*, Aug. 3, 1878, p. 2; *Philadelphia Inquirer*, Aug. 9, 1878, p. 8.

162 **"was more than usually unpropitious":** C. Abbe (1881:24).

162 **"Cold rain all this afternoon":** These telegrams are in WB NC-3, Entry 7, Box 342, Items 1171, 1176, 1184, 1197. Transcriptions are in C. Abbe (1881:30), with minor differences in wording.

163 **"the bluest party of astronomers":** *Chicago Times*, Aug. 3, 1878, p. 2.

163 **"I lay awake the second night":** *Pittsburgh Commercial Gazette*, Aug. 13, 1878, p. 1.

163 **"[W]e felt constant and severe headache":** USNO (1880:206).

163 **condition known today as high-altitude cerebral edema:** Hackett and Roach (2004).

164 **"otherwise known as 'Old Probabilities'":** *Denver Daily Times*, July 29, 1878, p. 4.

CHAPTER 14: FAVORED MORTALS

167 **"[I]t will probably be":** *Daily Graphic* (New York), July 29, 1878, p. 2 (vol. 17, p. 186).

167 **"in a manner never before possible":** *New York Herald*, July 29, 1878, p. 4.

167 THIS AFTERNOON'S ECLIPSE: *Sun* (New York), July 29, 1878, p. 1.

167 **"Professors Newcombe and Harkness":** *Philadelphia Inquirer*, July 29, 1878, p. 2.

168 **"First, the establishment":** *Chicago Times*, July 29, 1878, p. 5.

168 **"Should this body be discovered":** *Washington Post*, July 29, 1878, p. 1.

168 **"accident, the coming of disasters":** *Boston Daily Globe*, July 29, 1878, p. 1.

169 **"I could not help sitting down":** Schuster (1932:89).

169 **"[N]ot a cloud was to be seen":** From Trouvelot's official eclipse report in USNO (1880:75–94), which provides a detailed account of activities at Creston.

169 **"Everything promised well":** From Harkness's official eclipse report in USNO (1880:29–73), which is another important source of information on the day at Creston. See also *New York Herald*, July 30, 1878, p. 3 [OBSERVATIONS OF THE NAVAL OBSERVATORY PARTY IN WYOMING . . .], presumably written by Harkness.

169 **Down the tracks in Rawlins:** Most details of the events at Rawlins come from Edwin Marshall Fox's account in *The New York Herald*, July 30, 1878, p. 3. Also see Fox (1879:452).

169 **"with the force of a hurricane.":** *Standard* (London), Aug. 16, 1878, p. 2.

169 **James Craig Watson and Norman Lockyer:** Details of their activities can be found in several published accounts: Watson's official report in USNO (1880:117–24); Lockyer's unsigned article in *The Daily News* (London), Aug. 20, 1878, pp. 5–6 [also in *Nature* 18 (460), Aug. 22, 1878, pp. 430–32; the writer's identity is revealed by the *New-York Tribune*, Aug. 31, 1878, p. 3]; and Lockyer's signed account in *Nature* 18 (461), Aug. 29, 1878, pp. 457–62. Fred Hess, in Rawlins on eclipse day, wrote that "Prof. Watson and Mr. Lockyer left for Separation in order to gain about nine seconds of time, for observation of the total phase" ["SOL IN THE SHADE. Fred Hess Gives an Account of His Expedition," unidentified clipping presumed to be from *Webster County Gazette and Messenger* (Fort Dodge), Aug. 16, 1878, in USNO-NA Entry 18, Box 8].

170 **Joining them were several volunteers:** R. C. Lehmann wrote in his travel diary on July 29, 1878 [LFP Box 132, Folder 2]: "At 7:30 Lockyer, Professor & Mrs. Watson, Silvis the U.P.R.R. photographer, a young fellow called Talbot, Close and myself started after a hurried breakfast in the photographic car which was attached to a freight train."

170 **at six o'clock, he wrote: "Not a cloud":** *Chicago Times*, Aug. 3, 1878, p. 2.

170 **during the recent blue glass craze:** Azure glass found a new purpose across the country. *Chicago Daily Tribune*, July 20, 1878, p. 12: "Those who were victims of the blue-glass mania may find useful employment for that material at last." *Chicago Times*, July 30, 1878, p. 1: "[B]lue glass has been discovered to be the one thing above all others to look at a solar eclipse through." *Cedar Rapids Weekly Times*, Aug. 1, 1878, p. 3: "Persons who had invested in blue glass had an opportunity to reap the first benefits from their purchases."

170 **old stovepipe hats:** *Republican Daily Journal and Daily Kansas Tribune* (Lawrence), Aug. 4, 1878, p. 2.

171 **"Here's your eclipse glasses":** *Daily Rocky Mountain News* (Denver), July 30, 1878, p. 4.

171 **hill on the edge of the city:** Details of Maria Mitchell's activities are from Kendall (1896:226–32) [original in MMM Reel 4, Item 33]. Mitchell's viewing location was a tract known as McCullough's Addition. See *Daily Rocky Mountain News* (Denver), July 30, 1878, p. 4; *Daily Rocky Mountain News* (Denver), July 3, p. 4.

171 **the same telescope she had used:** Albertson (1913:9).

171 **St. Joseph's Home:** Fishell (1999:45), W. C. Jones and K. Forrest (1973:215).

172 **The city appeared to be on holiday:** The day's events were detailed in *The Denver Daily Tribune*, July 30, 1878, p. 4; *Daily Rocky Mountain News* (Denver), July 30, 1878, p. 4.

172 **sponsored by the Chicago Astronomical Society:** Its activities were recounted in Denver dailies and in Chicago Astronomical Society (1878).

172 **A rival team of Chicago astronomers:** *Chicago Times*, Aug. 3, 1878, pp. 1–2. For background on the rivalry, see *The Chicago Daily Tribune*, July 23, 1878, p. 8; *Chicago Times*, July 25, 1878, p. 4.

172 **evacuated the night before:** Cleveland Abbe wrote of his evacuation from the summit and his activities at the Lake House in C. Abbe (1881:24, 43, 46–48) and *Daily Gazette* (Colorado Springs), Aug. 7, 1878, p. 4 [reprinted in *Weekly Gazette* (Colorado Springs), Aug. 10, 1878, p. 5].

173 **"My Dear General":** Abbe to Myer, July 29, 1878 [WB NC-3, Entry 7, Box 359, Item 2054; also in C. Abbe (1881:31), with slight changes to wording and punctuation].

173 **"[A]t last we were among the favored":** *Georgetown Courier*, Aug. 1, 1878, p. 3.

174 **"Here ye are now":** *New York Herald*, July 30, 1878, p. 10.

174 **"Portly bankers about to start for home":** *New York Herald*, July 30, 1878, p. 10.

CHAPTER 15: FIRST CONTACT

175 **2:03:16.4 to 3:13:34.2 P.M.:** Time of first contact as observed by Simon Newcomb, that of second contact as reported by C. G. Bowman, both in USNO (1880:108).

175 **2:19:30.5 to 3:29:03.5 P.M.:** Times of first and second contact as reported by Elias

Colbert of the Chicago Astronomical Society party, *Chicago Daily Tribune*, July 30, 1878, p. 1.

175　**the moment he saw the moon kiss the sun:** *New York Herald*, July 30, 1878, p. 3.

175　**By now a crowd had gathered:** Among the onlookers was Ed Dickinson, division superintendent of the Union Pacific Railroad, who had graciously served as tour guide for the Draper party and was now joined by his family, from Laramie; see *Omaha Daily Bee*, July 29, 1878, p. 4.

176　**Soldiers from Fort Steele placed sections:** USNO (1880:111, 115, 118).

176　**Culbertson's nerves got the better of her:** Footnote by Phebe Mitchell Kendall appended to Maria Mitchell's account of events, in Kendall (1896:230). Mitchell does not identify the members of her party by name, but she does disclose their home states, which reveals which participant played which role on eclipse day.

176　**"Between first contact and totality":** MMM Reel 4, Item 33. Transcribed in Kendall (1896:230).

177　**two young men, who constructed a kind of easel:** C. Abbe (1881:46) and *Daily Gazette* (Colorado Springs), Aug. 7, 1878, p. 4 [reprinted in *Weekly Gazette* (Colorado Springs), Aug. 10, 1878, p. 5].

177　**dozens of tourists had arrived:** Mary Rose Smith penned an extended account of her group's activities on the peak for *The Philadelphia Inquirer*, Aug. 9, 1878, p. 8, which provides many of the details here. A brief report by a member of a second group of tourists is in *Illinois Daily State Journal* (Springfield), Aug. 2, 1878, p. 1.

177　**"no one not of the service":** *Daily Rocky Mountain News* (Denver), July 2, 1878, p. 4.

177　**Mrs. Smith and her small party took a seat:** Official reports by Mary Rose Smith (Mrs. Aubrey H. Smith) and two other members of the party, Edward Hatch and Charles B. Lamborn, are in C. Abbe (1881:39–41). Lamborn also shared his recollections in *The Daily Gazette* (Colorado Springs), July 31, 1878, p. 4 [reprinted in *Press* (Philadelphia), Aug. 6, 1878, p. 8].

178　**It squatted and grinned:** Smith's published report in *The Philadelphia Inquirer*, Aug. 9, 1878, p. 8, reads that the marmot "squalled" and grinned, an apparent typo. A clipping of this article submitted to Cleveland Abbe has been corrected—presumably by Smith herself—to read "squatted" and grinned [WB NC-3, Entry 7, Box 359, Item 2054].

178　**painting the shape of a sickle:** *Daily Rocky Mountain News* (Denver), July 30, 1878, p. 4.

179　**"a gorgeous appearance":** *Boulder County Courier*, Aug. 2, 1878, p. 3.

179　**called for a second jacket:** Ryan (1879:30).

179　**wrapped themselves in cloaks and blankets:** *Boulder County News*, Aug. 2, 1878, p. 3.

179　**"a very dark warm blue":** Penrose (1878a:189).

179　**"strange, weird, grey light":** *Standard* (London), Aug. 16, 1878, p. 2.

179　**"The light of the sun grew pale":** *Philadelphia Inquirer*, Aug. 9, 1878, p. 8.

179　**"which was shining brightly":** Ranyard (1881:219).

180　**owls emerged:** *Daily Rocky Mountain News* (Denver), Aug. 1, 1878, p. 4.

180　**"all the cocks in the city":** C. Abbe (1881:80).

180　**cows turned homeward:** *St. Louis Globe-Democrat*, July 30, 1878, p. 1; *Colorado Miner* (Georgetown), Aug. 3, 1878, p. 1.

180　**pigeons went to roost:** *Denver Daily Tribune*, July 30, 1878, p. 4; *New York Herald*, July 30, 1878, p. 3; *Colorado Weekly Chieftain* (Pueblo), Aug. 1, 1878, p. 3.

180   **Grasshoppers folded their wings:** *Cheyenne Daily Leader*, Aug. 3, 1878, p. 4; *Colorado Transcript* (Golden), Aug. 14, 1878, p. 3; C. Abbe (1881:80).

180   **the mayor stationed a police officer:** Chicago Astronomical Society (1878:13).

180   **"that if, during the totality":** *Laramie Daily Sentinel*, July 30, 1878, p. 4.

180   **"The only place of disorder":** *New York Herald*, July 30, 1878, p. 3.

180   **began sweeping the heavens:** USNO (1880:119), Watson (1878a:230–31).

181   **"The light seemed no longer":** USNO (1880:104).

181   **"They coursed after each other":** USNO (1880:152). Other accounts of shadow bands (sometimes called *diffraction* bands) can be found in USNO (1880:172, 176, 222, 224), Young (1878a:285).

181   **"children ran after it":** Airy (1851:65).

181   **the same that makes stars twinkle:** Zirker (1995:129–32).

181   **"visible wind":** Todd (1894:28).

182   **explained by British astronomer Francis Baily:** Baily (1836).

CHAPTER 16: TOTALITY

183   **3:13:34.2 to 3:16:24.2 P.M.:** Times of second and third contact as observed by C. G. Bowman, in USNO (1880:108).

183   **3:29:03.5 to 3:31:44.0 P.M.:** Times of second and third contact as reported by Elias Colbert of the Chicago Astronomical Society party, *Chicago Daily Tribune*, July 30, 1878, p. 1.

183   **"I was once held in a drowning condition":** Copernicus (1869:112).

183   **"The next morning, at the instant":** Arago (1858:362).

184   **"In fact, the general scene":** Smyth (1853:504–5).

184   **"darkness was like that of deep twilight":** *New York Herald*, July 30, 1878, p. 3.

184   **that job fell to Anna Draper:** Barker (1878:112); *Cheyenne Daily Leader*, July 31, 1878, p. 4. Years later, Cannon (1915a:381) wrote that Anna Draper had been assigned "to count the seconds during the eclipse and lest the vision might unnerve her, she was put within a tent and therefore saw nothing at all of the wonderful phenomenon." The claim, however, seems exaggerated. The Draper party possessed no tent, and in a letter penned in late 1878, Henry Draper suggested that his wife had gleaned at least "general impressions" of the eclipse despite her counting the seconds [Draper to Abbe, Dec. 7, 1878, CAP-JHU Box 1]. Anna Draper's assistance during the transit of Mercury is described in the *New-York Tribune*, May 7, 1878, p. 5.

184   **gave the duty to R. C. Lehmann:** R. C. Lehmann's travel diary, entry for July 29, 1878 [LFP Box 132, Folder 2].

184   **Bisbee hammered against the round cover:** USNO (1880:112).

185   **hid the brighter, inner corona behind a circular screen:** USNO (1880:103–4). Newcomb had tried the same technique during the 1869 eclipse in Des Moines but failed to execute the plan properly; see Newcomb (1869), Sands (1870:7–9).

185   **John W. Hoyt:** USNO (1880:103); H. S. Miller (1970:167–69); *Scientific American Supplement*, July 22, 1876, p. 469.

186   **"Was disappointed in not finding":** Newcomb diary, entry for Aug. 7, 1869 [SNP Box 1].

186   **announced the elapsed time. *Ten seconds*:** Exactly what Bisbee announced on every tenth beat is not clear. At this point in his counting, he may merely have said "ten," not "ten seconds." See more at USNO (1880:103).

186   **Watson pointed his telescope at the sun:** For details of Watson's sweep for Vulcan, see Watson (1878a), USNO (1880:119–20).

187   **"The most singular thing":** Airy (1852:32).

187   **"Totality had brought with it":** *New York Herald*, July 30, 1878, p. 3. See also Fox (1879:452), USNO (1880:376).

187   *phototelespectroscope:* Draper (1878b); *New-York Times*, Aug. 8, 1878, p. 5.

187   **"The amount of light":** Barker (1878:111).

187   **"were in excellent working order":** *New York Herald*, July 30, 1878, p. 3.

187   **a gap finer than the breadth of a hair:** *Daily Graphic* (New York), July 19, 1878, p. 1 (vol. 17, p. 121).

187   **Watson was making his fifth sweep:** Watson to John Rodgers, Aug. 13, 1878 [original in USNO-NA Entry 18, Box 8; published in *Astronomische Nachrichten* 93 (11), Sept. 22, 1878, pp. 161–64]; also *Chicago Times*, Aug. 6, 1878, p. 5.

188   **At almost the same moment:** Watson wrote that he found his mysterious object "about a minute before the end of the total eclipse" [*Detroit Post and Tribune*, Aug. 8, 1878, p. 1]. Edwin Marshall Fox noted that Edison finally concentrated the sun's rays on the rubber strip of his tasimeter "just as the chronometer indicated that but one minute remained of total eclipse" [*New York Herald*, July 30, 1878, p. 3].

188   **it shot off the scale entirely:** Edison (1879:54); *New York Herald*, July 30, 1878, p. 3; *World* (New York), Aug. 27, 1878, p. 1 [TAED SB032075a]; *Daily Graphic* (New York), Aug. 28, 1878, p. 3 (vol. 17, p. 399) [TAED MBSB10858].

188   **Edison was "in ecstasies":** *Standard* (London), Aug. 16, 1878, p. 2. A colorful story, told in later years, claimed that Edison's observations had been disrupted by chickens returning to the henhouse as darkness set in, but this tale appears apocryphal; see Eddy (1972:176–78).

189   **"It is of course now":** USNO (1880:105).

190   **"Those who have seen a locomotive approach . . .":** *Scientific American*, July 27, 1878, p. 49. The original quotation, in French, is in Forbes (1843:367).

190   **"a solid, palpable body":** C. Abbe (1881:41). Also in *Philadelphia Inquirer*, Aug. 9, 1878, p. 8.

190   **"an angry black cloud":** C. Abbe (1881:39).

190   **"cheer after cheer echoed":** *Denver Daily Tribune*, Aug. 2, 1878, p. 4.

190   **"[W]hen the total obscuration was reached":** *Chicago Times*, Aug. 3, 1878, p. 1.

190   **"It permeated everything":** *Chicago Times*, Aug. 3, 1878, p. 2.

190   **Perceptions of the corona are remarkably subjective:** See, e.g., USNO (1880:288): "Drawings of the corona made by different persons frequently bear so little resemblance to each other that it is difficult to believe them representations of the same thing. Even in the plainest details there are almost incredible differences of position and direction." Also Upton (1879:64): "It is singular that the corona should appear so differently to different observers, even if they are at the same station and view it under similar conditions."

191   **members of the class could not agree:** Chicago Astronomical Society (1878:11).

192   **"If I have the good fortune":** Ranyard (1881:235). See also *Chicago Times*, Aug. 9, 1878, p. 3.

192   **The women saw Mercury, Mars, and Venus:** Kendall (1896:232) [original in MMM Reel 4, Item 33]; *Scientific American*, Aug. 24, 1878, p. 113.

192   **"an attraction to the gaping":** *Inter Ocean* (Chicago), July 30, 1878, p. 1.

192 **"[W]omen of low and high degree"**: *Cincinnati Commercial*, Aug. 16, 1878, Supplement, p. 1. See also *Woman's Journal*, Aug. 17, 1878, p. 257.

193 **joined by Denver's bishop**: Ryan (1879:29–30).

193 **"Women are not needed as men"**: Machebeuf (1877:8).

195 **"I became perfectly convinced"**: C. Abbe (1881:47). See also *Daily Gazette* (Colorado Springs), Aug. 7, 1878, p. 4 [reprinted in *Weekly Gazette* (Colorado Springs), Aug. 10, 1878, p. 5].

195 **"No red color in corona"**: C. Abbe (1881:35).

195 **used an improvised *photometer***: USNO (1880:211–15).

195 **sketching the corona while viewing it with the naked eye**: USNO (1880:207–8).

195 **a secret wish—"to *see* the eclipse"**: Langley to Abbe, June 17, 1878 [CAP-LOC Box 8].

195 **"I once experienced an earthquake"**: *Pittsburgh Commercial Gazette*, Aug. 13, 1878, p. 1.

195 **Samuel Langley cried "over"**: C. Abbe (1881:39).

196 **"Then every tongue was unloosed"**: *Colorado Miner* (Georgetown), Aug. 3, 1878, p. 1.

196 **"The shadow! The shadow!"**: MMM Reel 4, Item 33. Transcribed in Kendall (1896:231–32).

196 **"looked like a black carpet"**: Freeman (1937:492).

196 **"a rounded ball of darkness"**: USNO (1880:216).

CHAPTER 17: AMERICAN GENIUS

197 **"[T]here were thousands of ignorant people"**: *St. Louis Globe-Democrat*, Aug. 6, 1878, p. 5.

197 **Chicago, Boston, Salt Lake City**: *Sunday Telegraph* (Chicago), Aug. 18, 1878, p. 3; *Chicago Daily Tribune*, Aug. 19, 1878, p. 5; *Boston Daily Globe*, Aug. 15, 1878, p. 2; *Boston Post*, Aug. 15, 1878, p. 4; *Salt Lake Daily Tribune*, Aug. 28, 1878, p. 1.

198 **"Professor Watson Finds Vulcan"**: *Laramie Daily Sentinel*, July 30, 1878, p. 4.

198 **"A new planet discovered"**: *Sioux City Daily Journal*, Aug. 2, 1878, p. 4.

198 **"[L]ittle doubt remains"**: *Daily News* (London), Aug. 20, 1878, p. 6 [also in *Nature* 18 (460), Aug. 22, 1878, p. 432]. Also see Lockyer interview in *Washington Post*, Aug. 9, 1878, p. 1.

198 **in private conversation, questioned Watson's discovery**: S. W. Burnham to Edward S. Holden, Aug. 12, 1878 [USNO-NA Entry 18, Box 8]: "I wonder if Watson really saw anything but a star." Holden to C. H. F. Peters, Aug. 31, 1878 [CHFPP Box 0000.123.2.3]: "I talked with Watson—he seems confident of his observation, but I can't help thinking he might have mistaken a star for his planet." Watson quoted in *Chicago Times*, Aug. 6, 1878, p. 5: "Some of the other astronomers out in the mountains were skeptical as to my discovery, although most of them have since become convinced."

198 **"a thoroughly competent and trustworthy observer"**: *Boston Daily Advertiser*, July 31, 1878, p. 2.

198 **"the most noted astronomical observer"**: *Laramie Daily Sentinel*, July 30, 1878, p. 4.

198 **"He is the Edison of astronomy"**: *Cheyenne Weekly Leader*, Oct. 3, 1878, p. 4.

198 **"What were the circumstances"**: *Chicago Times*, Aug. 6, 1878, p. 5.

199 **a handicap—called a** *discount*: Phelan (1858:68).

199 **"The success of this party":** *Sun* (New York), Aug. 4, 1878, p. 2.

199 **"Recently, here in our midst":** *Daily Rocky Mountain News* (Denver), Aug. 4, 1878, p. 2.

199 **"the most distinguished lady astronomer of the age":** *Daily Rocky Mountain News* (Denver), July 28, 1878, p. 1.

200 **"when [the king] found that she was doing":** This quotation comes from a previous lecture [*New-York Tribune*, Jan. 15, 1873, p. 4], but Mitchell likely made a similar remark in Denver. For more on Caroline Herschel, see Hoskin (2011).

200 **"were as peculiar and as plainly marked":** *Daily Rocky Mountain News* (Denver), July 31, 1878, p. 4. She made the same points, using similar language, in M. Mitchell (1889:908).

200 **an elaborate flower arrangement:** *Denver Daily Tribune*, July 31, 1878, p. 4; *Denver Daily Times*, July 31, 1878, p. 4.

200 **symbolizing sunrise through the pall:** Ellis (1961:17).

201 **"The observation of the eclipse":** *New York Herald*, July 30, 1878, p. 3.

201 **"Edison's experiments with the tasimeter":** *New-York Times*, July 30, 1878, p. 5.

201 **"Edison's trip to the west":** *Evening Star* (Washington, D.C.), Aug. 1, 1878, p. 2.

201 **"His plan is to adjust his tasimeter":** *New York Herald*, Aug. 1, 1878, p. 7.

202 *Scientific American* **praised the concept:** *Scientific American*, Aug. 24, 1878, p. 112 [TAED MBSB10830X].

202 **"The meeting here, will I fear":** Newcomb to his wife, Aug. 21, 1878 [SNP Box 9].

202 **"Edison is undoubtedly the 'big gun'":** *St. Louis Globe-Democrat*, Aug. 24, 1878, p. 2.

203 **"I haven't written the paper yet":** *St. Louis Evening Post*, Aug. 22, 1878, p. 4. George Barker helped Edison prepare for the talk. See Barker to Edison, July 12, 1878 [TAED D7802ZSG; also in TAEB vol. 4, p. 393], and lecture notes in Barker's handwriting, Aug. 20, 1878 [TAED D7835L].

203 **"I believe that is all":** *Missouri Republican* (St. Louis), Aug. 24, 1878, p. 5.

203 **"reflected glory upon the progressive genius":** *St. Louis Evening Post*, Aug. 23, 1878, p. 1.

203 **THE GREAT EDISON. THE SCIENTISTS WELCOME HIM:** *Boston Post*, Aug. 24, 1878, p. 2. For more on Edison's attendance at AAAS, see *Missouri Republican* (St. Louis), Aug. 23, 1878, p. 8; *St. Louis Evening Post*, Aug. 24, 1878, p. 3; *New-York Tribune*, Aug. 24, 1878, p. 2 [TAED SB032055c; also in TAEB vol. 4, pp. 439–41]; *Chicago Daily Tribune*, Aug. 24, 1878, p. 3; *Cincinnati Daily Gazette*, Aug. 24, 1878, p. 1; *Sunday Telegraph* (Chicago), Aug. 25, 1878, p. 5 [TAED SB032088c]; *New-York Tribune*, Aug. 26, 1878, p. 5; AAAS (1879).

203 **"Why, Dot, is that you?":** *World* (New York), Aug. 27, 1878, p. 1 [TAED SB032075a].

203 **Several journalists had come:** *Daily Graphic* (New York), Aug. 28, 1878, p. 3 (vol. 17, p. 399) [TAED MBSB10858]; *Sun* (New York), Aug. 29, 1878, p. 3 [TAED SB032060a]; *New York Herald*, Aug. 29, 1878, p. 4 [TAED SB032059a].

203 **"Did you get any new ideas":** *World* (New York), Aug. 27, 1878, p. 1 [TAED SB032075a].

CHAPTER 18: GHOSTS

207   **the era that created the America we recognize:** Nevins (1927), Goldfield (2011).

207   **outperforming Britain, France, and Germany:** Bairoch (1982:296).

207   **"It must be acknowledged that":** Tocqueville (1863:40).

208   **"In other intellectual nations":** Newcomb (1876:118).

208   **"It is good for our general culture":** *Christian Union*, April 17, 1878, p. 328.

208   **"When the time of our late eclipse":** *Inter Ocean* (Chicago), Dec. 2, 1878, p. 2; also in Swing (1889:19–21). More at Newton (1909).

209   **"Colorado has . . . furnished":** *Colorado Banner* (Boulder), Aug. 1, 1878, p. 1.

209   **"It is thought that the results":** John Rodgers to R. W. Thompson, Dec. 12, 1878 [USNO-NA Entry 2, vol. 3, p. 567].

209   **General Myer assigned Cleveland Abbe:** Myer issued his order on Aug. 13, 1878 [WB NC-3, Entry 43, vol. 6, p. 187].

209   **a theory he had devised just after totality:** C. Abbe (1881:48); *Daily Gazette* (Colorado Springs), Aug. 7, 1878, p. 4 [reprinted in *Weekly Gazette* (Colorado Springs), Aug. 10, 1878, p. 5].

209   **a full 461 pages:** Abbe submitted his report on Oct. 17, 1878 [WB NC-3, Entry 8, vol. 9, pp. 376–77; a note in the margin reads "One enclosure (461 pages . . .)"].

210   **put a stop to the whole project:** C. Abbe (1881:3).

210   **"Was ever [an] astronomer":** *Chicago Daily Tribune*, Aug. 15, 1878, p. 4.

210   **periodicals in London:** *Echo* (London), Aug. 30, 1878, p. 2; *Cornhill*, Dec. 1878, pp. 714–22.

210   **"I regard this eclipse":** *Nature* 18 (461), Aug. 29, 1878, p. 460.

210   **"which American men, and American instruments":** Smyth to Draper, Sept. 27, 1878 [HMAPDP].

211   **Woman's Congress, this time in Providence:** For general information on the meeting, see *Providence Daily Journal*, Oct. 10, 1878, p. 1 and p. 2; *Boston Daily Globe*, Oct. 10, 1878, p. 2; *Providence Daily Journal*, Oct. 11, 1878, p. 1; *Woman's Journal*, Oct. 19, 1878; AAW (1879). Ticket sales reported in AAW (1880).

211   **Mitchell ascended and drew three circles in chalk:** *Providence Daily Journal*, Oct. 12, 1878, p. 1; *Woman's Words*, Nov. 1878, p. 300.

211   **"In a total eclipse of the sun":** All quotations are from Mitchell's original lecture notes [MMM Reel 4, Item 33]. The speech also appears in Kendall (1896:223–32), in slightly edited form.

212   **"not the flitting of the cloud shadow":** In transcribing her sister's handwriting, Kendall (1896:232) interpreted this phrase to read "not the flitting of the *closer* shadow."

212   **"conclusive proof":** *Inter Ocean* (Chicago), Oct. 26, 1878, p. 9.

212   **"Professor Maria Mitchell of Vassar":** *Boston Daily Advertiser*, Oct. 15, 1878, p. 2.

212   **"by the clanging chains":** Thomas (1908:49).

213   **Dr. Emma Culbertson:** Although not a founding member of the Association of Collegiate Alumnae, Culbertson was active in the organization from its early days. See *Vassar Miscellany*, Jan. 1885, p. 1207; Talbot and Rosenberry (1931:125–26).

213   **"The graduates, as a body":** Howes et al. (1885:77). See more at Zschoche (1989:563).

213   **"If I am to get really well":** Mitchell diary entry for May 30, 1881 [MMM Reel 3, Item 24, p. 78; transcribed in Booker (2007:430)].

214 **"brilliant discovery"**: French astronomer Félix Tisserand saluted Watson's "brillante découverte de Vulcain" in a letter dated Sept. 2, 1878. German astronomer C. A. F. Peters (not to be confused with the American C. H. F. Peters) sent his congratulations on Sept. 24, 1878. [Both letters are in JCWP Box 1.]

214 **"Permit me to congratulate you"**: Bell to Watson, Dec. 28, 1878 [AGBFP Box 304].

214 **"[H]e will be a most valuable addition"**: *Wisconsin Journal of Education*, Nov. 1878, p. 525. For more on Watson's move to Wisconsin and the University of Michigan's attempts to forestall it, see *Detroit Free Press*, Oct. 5, 1878, p. 4; *Detroit Free Press*, Oct. 12, 1878, p. 4 and p. 8; *Detroit Free Press*, Oct. 13, 1878, p. 8; *Chicago Daily Tribune*, Aug. 6, 1879, p. 7; N. B. Van Slyke to Watson, Oct. 9, 1878 [JCWP Box 1].

214 **The second charmed observer was Lewis Swift**: Swift is best remembered today as co-discoverer of Comet Swift-Tuttle, which produces a trail of dust that rains down on earth each August as the Perseid meteor shower. See Wlasuk (1996).

214 **he, like Watson, had spied an unknown object**: *Chicago Daily Tribune*, Aug. 2, 1878, p. 3; *Buffalo Daily Courier*, Aug. 6, 1878, p. 1; *New-York Times*, Aug. 16, 1878, p. 5; *New-York Tribune*, Aug. 21, 1878, p. 1; *New-York Tribune*, Aug. 22, 1878, p. 5; *Scientific American*, Aug. 31, 1878, p. 128; *Chicago Daily Tribune*, Sept. 10, 1878, p. 3; *Nature* 18 (464), Sept. 19, 1878, p. 539.

215 **"I have no doubt about"**: Watson to John Rodgers, Sept. 2, 1878 [USNO-NA Entry 7, Box 45; copy in JCWP Box 2]. Watson made a similar statement to USNO astronomer Asaph Hall in a letter dated Aug. 21, 1878 [AHP Box 3].

215 **"[T]he records of my circles"**: Watson (1878*b*:312).

215 **"This Vulcan business"**: *Inter Ocean* (Chicago), Sept. 16, 1878, p. 4.

215 **"If the above conclusions are true"**: *Nature* 19 (475), Dec. 5, 1878, p. 96. See also *Rochester Democrat and Chronicle*, Dec. 19, 1878, p. 2.

215 **an even more derogatory title**: Asim (2007:103–4).

215 **"But," Newcombe wrote, "the Vulcans have increased by one"**: Newcomb to Peters, May 14, 1879 [SNP Box 4].

216 **"One little Vulcan"**: Newcomb wrote, in German: "Ein kleine Vulcan isst heissen Brei/ Er sprang in die Höhe, und Siehe!/ es sind zwei./ zwei kleine Vulcan etc./ und siehe es sind drei."

216 **"I should have liked to hear"**: Peters to Newcomb, May 18, 1879 [SNP Box 35].

216 **"Said one to the other"**: Peters wrote, in a mixture of German and English: "Sprach einer zum andern: 'let's drink little beer!'/ Sie tranken recht lustig,—und sieh' es sind *vier*!/ Sie möchten wohl gerne einen *fünften* zur Noth,/ Doch's Fass ist ja leer,—sie tranken sich todt./ To which may be added perhaps the lesson:/ So geht's, wenn der Frosch sich blähet zu sehr;/ Er platzt doch am Ende,—ihm glaubt keiner mehr."

217 **"in the dim light of the total eclipse"**: Peters (1879).

217 **"Professor Peters' whole attack"**: Watson (1879). See also Watson to C. H. F. Peters, May 15, 1879 [Vertical File Collection, DB–James C. Watson, Bentley Historical Library, University of Michigan].

217 **Since the age of Aristotle, a myth has persisted**: D. W. Hughes (1983).

218 **specially designed to look for Vulcan**: Watson (1880); *Wisconsin State Journal* (Madison), Sept. 28, 1880, p. 8; *Scientific American*, Dec. 25, 1880, p. 405; Comstock (1895:52). Watson explained the basis for his plan immediately after returning from Wyoming; see *Detroit Free Press*, Aug. 8, 1878, p. 4.

218 **"I am happy to say"**: Watson to Julius E. Hilgard, Nov. 15, 1880 [GADRUWM Series 7/4/3].

218 **"Look out for a 'freeze up'":** *Galesville Independent*, Nov. 11, 1880, p. 3.

218 **"out among the workmen":** Hazen Mooers to Mrs. James C. Watson, Dec. 27, 1880 [JCWP Box 1].

218 **Death arrived for the astronomer:** *Milwaukee Daily Sentinel*, Nov. 24, 1880, p. 1; *Evening News* (Detroit), Nov. 24, 1880, p. 1; *Detroit Post and Tribune*, Nov. 24, 1880, p. 1; *Wisconsin State Journal* (Madison), Nov. 30, 1880, p. 5.

218 **"recklessness of his own health":** *Chronicle* (Univ. of Michigan), Nov. 27, 1880, p. 65.

218 **"exhausted condition by overwork and exposure":** *Inter Ocean* (Chicago), Nov. 24, 1880, p. 2.

218 **"The Most Brilliant":** *Chicago Daily Tribune*, Nov. 24, 1880, p. 2.

219 **august memorial service:** *Inter Ocean* (Chicago), Nov. 27, 1880, p. 6; *Detroit Post and Tribune*, Nov. 27, 1880, p. 1; *Chronicle* (Univ. of Michigan), Nov. 27, 1880, pp. 65–66; *Ann Arbor Register*, Dec. 1, 1880, pp. 1–2.

219 **"the brightest scientific ornament":** Frieze et al. (1882:25).

219 **"with what delight":** Frieze et al. (1882:17).

219 **"One thing I would beg":** Peters to Holden, Feb. 17, 1881 [LOR Box 36, Folder 1].

219 **"No stars were seen":** Holden (1882:36).

219 **"abandoned as entirely useless":** *Sunday Republican-Sentinel* (Milwaukee), Oct. 29, 1882, p. 7.

220 **In 1919, Einstein's abstruse theory was famously put to the test:** Dyson, Eddington, and Davidson (1920). See more at Crelinsten (2006), Westfall and Sheehan (2015:138–41).

220 **"[W]hile it demonstrates his large love of science":** *Inter Ocean* (Chicago), Nov. 29, 1880, p. 4.

220 **a perpetual fund for two purposes:** National Academy of Sciences (1909:51–52). For more on how the money was spent in tracking Watson's asteroids, see *Boston Daily Globe*, Nov. 15, 1941, p. 2; Coddington (1900); Hansen (1943). The academy also granted $500 from the Watson Fund to aid in a search for Vulcan during the May 6, 1883, total solar eclipse; True (1913:64).

220 **his ego would endure:** Simon Newcomb, in a letter to the newly widowed Annette Watson, praised her late husband's generosity for funding a prize that would "carry his name to remote generations as a benefactor of young astronomers seeking recognition of their labors" [Newcomb to Mrs. Watson, Dec. 8, 1880, SNP Box 5]. Watson's brother, however, wrote harshly in a letter to the editor: "Professor Watson was a man of extraordinary intellect; but of his inordinate ambition to distinguish himself before all moral obligation to his family, there can be no better proof than his will. He gave all he possessed to a scientific society, with a view evidently, of still further perpetuating his name." [*Daily Pioneer Press* (St. Paul), Dec. 17, 1880, p. 5]

CHAPTER 19: SHADOW AND LIGHT

221 **one of the most celebrated feats in the history of technology:** For details of Edison's race to build his incandescent lamp, and of earlier attempts at electric illumination, see Edison (1904); Friedel, Israel, and Finn (2010); Freeberg (2013); Jonnes (2003).

221 **Avenue de l'Opéra:** *St. Louis Globe-Democrat*, Aug. 18, 1878, p. 11. Edison had apparently seen this article [TAED D7819C].

222 **"My dear Edison"**: Barker to Edison, Sept. 2, 1878 [TAED D7802ZXS].

222 **spoken at length about electric power and lighting**: TAEB vol. 4, p. 866; Barker's testimony in *Edison v. Siemens v. Field*, pp. 178–85 [TAED QD001178].

222 **"fairly gloated over it"**: *Sun* (New York), Sept. 10, 1878, p. 1.

222 **"I have it now!"**: *Sun* (New York), Sept. 16, 1878, p. 3 [TAED MBSB20887; also in TAEB vol. 4, pp. 503–6].

223 **"When do you expect"**: *New York Herald*, Oct. 12, 1878, p. 4 [TAED MBSB20949].

223 **Edison received inquiries**: William F. Barrett to Edison, Aug. 26, 1878 [TAED D7802ZWL; also in TAEB vol. 4, pp. 442–43]; William Curtis Taylor to Edison, Sept. 2, 5, and 7, 1878 [TAED D7835M, D7835N, and D7835O]; Jonathan Chace to Edison, Sept. 7, 1878 [TAED D7835P]; Joseph Tingley to Edison, Dec. 13, 1878, and Feb. 5, 1879 [TAED D7835T and D7935B].

223 **"it is of no interest"**: *Inter Ocean* (Chicago), July 16, 1878, p. 8.

223 **he permitted two companies**: These were the British instrument maker John Browning and the American firm of Partrick & Carter.

223 **without paying him royalties**: Edison to Browning, Sept. 19, 1878 [TAED LB003376].

223 **"The movements of the needle"**: *English Mechanic and World of Science*, Sept. 13, 1878, p. 13.

223 **"the instrument has been generally"**: Mendenhall (1882:43).

223 **"Edison's tasimeter was like"**: *Silver World* (Lake City, Colorado), Aug. 10, 1878, p. 2.

223 **"My apparatus was entirely too sensitive"**: Dyer and Martin (1910:230).

224 **People began to wonder if the Wizard was a sham**: *Daily Graphic* (New York), Feb. 6, 1879, p. 3 (vol. 18, p. 655) [TAED MBSB21122X]; *Scientific American*, Feb. 15, 1879, p. 102.

224 **he would underwrite the launch**: Anonymous (1947), Baldwin (1995:120–22).

224 **"[K]eep yourself aloof"**: Fox to Edison, Oct. 20, 1878 [TAED D7805ZDW; also in TAEB vol. 4, pp. 630–32].

225 **$125 loan**: Fox to Edison, May 29, 1879 [TAED D7903ZEK; also in TAEB vol. 5, pp. 247–48].

225 **gift of eight shares**: Calvin Goddard to Edison, Jan. 22, 1879 [TAED D7920M; also in TAEB vol. 5, pp. 44–46]; Fox to Edison, Jan. 26, 1879 [TAED D7920R].

225 **"The Great Inventor's Triumph"**: *New York Herald*, Dec. 21, 1879, p. 5 [TAED MBSB21379X].

225 **praised the inventor as "no unwary experimenter"**: *Nature* 18 (459), Aug. 15, 1878, p. 402.

225 **"absolute incompatibility with"**: *Nature* 21 (537), Feb. 12, 1880, p. 342.

225 **expressed himself "under so many obligations"**: Morton to Edison, Feb. 5, 1878 [TAED D7802ZAH].

225 **"a conspicuous failure"**: Morton letter to the editor of *The Plumber and Sanitary Engineer*, published Jan. 1, 1880, pp. 45–46 [TAED SM051037a]. Also in *Sun* (New York), Dec. 27, 1879, p. 3; *New York Herald*, Dec. 28, 1879, p. 7 [TAED MBSB21396X].

225 **"a lamp which surpasses"**: *Evening Post* (New York), Nov. 22, 1880, p. 3 [TAED MBSB21545X].

226 **"When I am through with my light"**: *Sun* (New York), Jan. 30, 1880, p. 2.

226 **electric lathes and clocks, electric hairbrushes**: *Times* (London), Aug. 11,

1881, p. 10; *Cincinnati Daily Gazette*, Aug. 25, 1881, p. 3; *Cleveland Herald*, Nov. 23, 1881, p. 3.

226   **in two opulent rooms:** *Daily News* (London), Aug. 11, 1881, pp. 5–6; *New York Herald*, Aug. 20, 1881, p. 7.

226   **his many newfangled contraptions:** Among the devices on display was one meant to demonstrate the etheric force, the existence of which Edison still occasionally advocated. See Dyer and Martin (1910:579); Simonds (1934:206); *Daily Graphic* (New York), Dec. 28, 1878, p. 6 (vol. 18, p. 394) [TAED MBSB21091]; *New York Herald*, Dec. 28, 1879, p. 7 [TAED MBSB21396X].

227   **"Surely the game of 'throwing light'":** *New-York Tribune*, Sept. 19, 1881, p. 5.

227   **"In the matter of the diminutive":** *Cleveland Leader*, Oct. 20, 1881, p. 6.

227   **"filled the hall with anthems":** *Evening Star* (Washington, D.C.), Nov. 23, 1881, p. 5.

228   **"Accept my congratulations":** *New-York Tribune*, Oct. 28, 1881, p. 5. A handwritten copy of Barker's cable is in Edison's letter book [TAED LM001069C].

228   **"It is to Mr. T. A. Edison":** Morton (1889:189).

228   **pressed to have his friend elected:** Barker to Newcomb, Dec. 4, 1908 [SNP Box 15].

229   **"This electrical age":** *New York Times*, Oct. 19, 1931, p. 20.

229   **synchronized minute in the dark:** *New York Times*, Oct. 21, 1931, p. 1.

229   **Albert Einstein and Pope Pius XI:** *New York Times*, Oct. 19, 1931, p. 26 and p. 1.

229   **forty thousand mourners:** *New York Times*, Oct. 21, 1931, p. 3. See also *New York Times*, Oct. 20, 1931, p. 1 and p. 3.

229   **a trove of mementoes:** Edison's library has been preserved within the Thomas Edison National Historical Park, and the room has been kept decorated much as it was at the time of the inventor's death. For a description of the library's history and furnishings, see Millard, Hay, and Grassick (1995).

EPILOGUE: TENDRILS OF HISTORY

231   **the historical marker tells a story:** "THOMAS A. EDISON CAMPED NEAR THIS SPOT IN 1878, WHILE ON A FISHING TRIP. IT WAS HERE THAT HIS ATTENTION WAS DIRECTED TO THE FIBER FROM HIS BAMBOO FISHING POLE WHICH HE TESTED AS A SUITABLE FILAMENT FOR HIS INCANDESCENT ELECTRIC LAMP."

231   **"After we had been [at Battle Lake] about three days":** Galbraith (1922:25). This story was retold many times by Galbraith; see *Omaha World-Herald*, Oct. 29, 1929, p. 7; *The Commonwealther*, Christmas 1929, pp. 20–21; *Arkansas Gazette* (Little Rock), Nov. 1, 1931, Features Section, p. 1 and p. 12; Galbraith to George E. Howard, Nov. 10, 1929 [TAEC Series XVII Unprocessed]. Galbraith's account is colorfully vivid, yet it contains several blatant inaccuracies that undermine its credibility. Galbraith claimed, for instance, that Edwin Marshall Fox had been on the fishing trip and promptly filed a dispatch about Edison's idea for an incandescent lamp, only to have *The New York Herald* reject the story as "rot." Fox was not, however, at Battle Lake with Edison, but rather was in Vermont with his dying brother; see *Boston Post*, Aug. 19, 1878, p. 2. Galbraith also recalled that the Meeker Massacre occurred during the fishing trip and when news of the incident reached Battle Lake, Major Thornburgh

was summoned to the Ute Reservation, where he was soon killed. Galbraith, however, has the chronology all wrong. Those violent incidents did not take place until September 1879—more than a year after Edison departed Rawlins— and they transpired in a different order. Interestingly, though, Thornburgh was indeed on a hunting and fishing trip near Battle Lake when the troubles began; see Sprague (1957:180). For more on the questionability of Galbraith's story, see Roberts (1981).

231 **Later embellishments:** Spring (1968), Steel (2001:207).

232 **relaxed and ready to take on new projects:** George Barker to Stockton Griffin, Sept. 5, 1878 [TAED D7819D; also in TAEB vol. 4, pp. 468–69]; *New York Herald*, Dec. 21, 1879, p. 5 [TAED MBSB21379X]; F. A. Jones (1908:97).

232 **"We often hear it, for instance":** Langley (1889:2–3).

233 **"[O]ur traditional reputation":** Newcomb (1897:712).

233 **"[T]o-day our country stands":** Newcomb (1897:714).

233 **Modern historians generally agree:** Reingold (1972:55), Rothenberg (1986:123), Moyer (1992:183–84).

234 **"the leader of the world in science":** Newcomb (1876:117).

234 **"Not only is our scientific literature":** Newcomb (1876:110).

234 **his meeting Norman Lockyer:** Speiden (1947:141).

234 **"It cost me too much":** TAEB vol. 2, p. 778.

234 **resurrected by Alexander Graham Bell:** Bruce (1973:376–78).

234 **"The precise form":** Newcomb (1876:118).

235 **endowed a second professional prize:** True (1913:365–66).

235 **"We are deficient":** Newcomb (1876:118).

235 **Harvard gave the work to an all-female team:** Reed (1892); B. Z. Jones and L. G. Boyd (1971:211–45); Mack (1990); Sobel (2016); *Boston Herald*, Jan. 16, 1899, pp. 1–2.

235 **"a corps of lady assistants":** Daniel Draper to Mary Anna Palmer Draper, May 12, 1885 [JWDFP Box 42].

236 **"I knew about Maria Mitchell":** Lightman and Brawer (1990:287). See also Rubin (1997:153, 164–72).

236 **"It is, in fact, thought":** *Times* (London), July 29, 1878, p. 11.

236 **astronomers head to the ends of the earth:** Pasachoff (2009).

AFTERWORD: UNISON

239 **the musical *American Eclipse*:** The piece was written by composer/lyricist Michael John LaChiusa.

240 **"Jay Pasachoff, Who Pursued Eclipses":** https://www.nytimes.com/2022/11/20/obituaries/jay-pasachoff-dead.html.

240 **"first all-American total solar eclipse":** *Nature*, May 25, 2017, p. 409.

241 **"[W]e're trying to persuade":** *Quanta Magazine*, Aug. 10, 2017.

241 **"Take a kid to the eclipse":** *New York Times*, Aug. 15, 2017, p. D5.

241 **"[T]he most important scientific outcome":** *Washington Post*, Aug. 13, 2017, p. A21.

241 **"It just shows us how powerful":** NBC News live coverage, Aug. 21, 2017. Accessed at https://www.youtube.com/watch?v=nK9O5sNuVL8. Quote is at 2:34.

242   **"Given the experiences we've had":** *North Jersey Herald News*, Aug. 22, 2017.

242   **"[I]t's immaculate":** *The Bulletin* (Bend, Oregon), Aug. 24, 2017.

242   **"Nation united in awe":** https://www.ajc.com/news/nation-united-awe/GgxIoVYBKyuGd5BZQUUXaM/.

242   **"This is a level of exposure":** Miller, Jon D. 2018, "Americans and the 2017 Eclipse: A final report on public viewing of the August total solar eclipse," Univ. of Michigan. Accessed at https://isr.umich.edu/wp-content/uploads/2018/08/Final-Eclipse-Viewing-Report.pdf.

242   **"exhibited more awe":** Goldy, Sean P., et al. 2022, "The Social Effects of an Awesome Solar Eclipse." *Psychological Science* 33 (9):1452–1462.

243   **"Before you die":** David Baron at TEDxMileHigh, July 8, 2017, Online at https://www.ted.com/talks/david_baron_you_owe_it_to_yourself_to_experience_a_total_solar_eclipse.

LIST OF ILLUSTRATIONS

The illustrations in this work derive from nineteenth-century books, magazines, newspapers, and archival collections. Most of the captions are presented as originally written, although they have been re-typeset for legibility and occasionally edited for length or spelling. See "Notes on Sources" for abbreviations used here.

FRONTISPIECE "The Great Solar Eclipse—Sketched at Snake River Pass, Colorado, by St. George Stanley," from *Harper's Weekly*, August 24, 1878. Author collection. Inset: Maria Mitchell, from *The Century Magazine*, October 1889; Thomas A. Edison, from *The Popular Science Monthly*, August 1878; James Craig Watson, from *The Popular Science Monthly*, September 1881. Courtesy of Denver Public Library.

ART INSERT

SELECT BIBLIOGRAPHY

Aaboe, A., et al. 1991. "Saros Cycle Dates and Related Babylonian Astronomical Texts." *Transactions of the American Philosophical Society* 81 (6):1–75.

Abbe, Cleveland. 1869. "The Resuscitation of the Cincinnati Observatory." In *Proceedings of the American Association for the Advancement of Science, Seventeenth Meeting, Held at Chicago, Illinois, August, 1868*, 172–74. Cambridge, Mass.: Joseph Lovering.

———. 1871. "Historical Notes on the Systems of Weather Telegraphy, and Especially Their Development in the United States." *American Journal of Science and Arts* [Third Series] 2 (8):81–88.

———. 1872 "Observations on the Total Eclipse of 1869." *American Journal of Science and Arts* [Third Series] 3 (16):264–67.

———. 1881. *Report on the Solar Eclipse of July, 1878*. Professional Papers of the Signal Service, no. 1. Washington, D.C.: Government Printing Office.

———. 1895. "The Meteorological Work of the U.S. Signal Service, 1870 to 1891." In *U.S. Department of Agriculture Weather Bureau, Bulletin No. 11: Report of the International Meteorological Congress, Held at Chicago, Ill., August 21–24, 1893, under the Auspices of the Congress Auxiliary of the World's Columbian Exposition*, Part 2, edited by Oliver L. Fassig, 232–85. Washington, D.C.: Weather Bureau.

———. 1916. "A Short Account of the Circumstances Attending the Inception of Weather Forecast Work by the United States." *Monthly Weather Review* 44 (4):206–7.

Abbe, Truman. 1955. *Professor Abbe and the Isobars: The Story of Cleveland Abbe, America's First Weatherman*. New York: Vantage.

Abbott, Frances M. 1889. "Maria Mitchell at Vassar." *Wide Awake* 29 (4):242–47.

Académie des Sciences. 1846. "ASTRONOMIE.—Planète de M. Le Verrier." *Comptes Rendus Hebdomadaires des Séances de l'Académie des Sciences* 23 (October 5):659–63.

Adams, Charles Francis, ed. 1874. *Memoirs of John Quincy Adams, Comprising Portions of His Diary from 1795 to 1848*. Vol. 1. Philadelphia: J. B. Lippincott.

Adams, W. G. 1871. "Report of Professor W. G. Adams, on Observations of the Eclipse of December 22nd, 1870, Made at Augusta, in Sicily." *Monthly Notices of the Royal Astronomical Society* 31 (5):155–61.

Airy, George B. 1851. "On the Total Solar Eclipse of 1851, July 28." *Notices of the Proceedings at the Meetings of the Members of the Royal Institution* 1 (1851–54):62–68.

———, ed. 1852. *Memoirs of the Royal Astronomical Society* 21, Part 1: *Containing the Observations of the Total Solar Eclipse of July 28, 1851*. London: Royal Astronomical Society.

Albers, Henry, ed. 2001. *Maria Mitchell: A Life in Journals and Letters*. Clinton Corners, N.Y.: College Avenue Press.

Albertson, Mary A. 1913. "Curator's Report." In *Annual Report of the Maria Mitchell Association*, vol. 11, 9–12. Nantucket, Mass.: Nantucket Maria Mitchell Assoc.

Ambrose, Stephen E. 2000. *Nothing Like It in the World: The Men Who Built the Transcontinental Railroad 1863–1869*. New York: Simon and Schuster.

American Association for the Advancement of Science. 1878. *Daily Programme of the Twenty-Seventh Meeting of the American Association for the Advancement of Science. Commencing Wednesday, August 21st, 1878. At St. Louis, Missouri*. St. Louis: Local Committee of the AAAS.

———. 1879. *Proceedings of the American Association for the Advancement of Science, Twenty-Seventh Meeting, Held at St. Louis, Missouri, August, 1878*. Salem, Mass.: AAAS.

Anonymous. 1876. *Historical Sketch of Vassar College. Founded at Poughkeepsie, New York, January 18, 1861*. New York: S. W. Green.

———. 1878a. "Edison's Micro-Tasimeter." *Journal of the Franklin Institute* 106 (633):173–76.

———. 1878b. "Edison's Tasimeter." *Manufacturer and Builder* 10 (7):147–48.

———. 1878c. "The Late Total Eclipse of the Sun, as Seen from Denver." *The Friend* 52 (16):121–22.

———. 1892. *A Memorial and Biographical History of Johnson and Hill Counties, Texas*. Chicago: Lewis Publishing.

———. 1947. "Thomas A. Edison and the Founding of *Science*: 1880." *Science* [New Series] 105 (2719):142–48.

Arago, François. 1858. *Popular Astronomy*. Translated and edited by Admiral W. H. Smyth and Robert Grant. Vol. 2. London: Longman, Brown, Green, Longmans, and Roberts.

Armitage, Angus. 1966. *Edmond Halley*. London: Thomas Nelson.

Armitage, Geoff. 1997. *The Shadow of the Moon: British Solar Eclipse Mapping in the Eighteenth Century*. Tring, Hertfordshire, U.K.: Map Collector Publications.

Armstrong, Henry E. 1928. "Norman Lockyer's Work and Influence." *Nature* 122 (3084):870–74.

Ashbrook, Joseph. 1973a. "The Adventures of C. H. F. Peters—I." *Sky and Telescope* 45 (2):90–91.

———. 1973b. "The Adventures of C. H. F. Peters—II." *Sky and Telescope* 45 (3):152–53.

Asim, Jabari. 2007. *The N Word: Who Can Say It, Who Shouldn't, and Why*. Boston: Houghton Mifflin.

Association for the Advancement of Woman. 1877. *Papers Read at the Fourth Congress of Women, Held at St. George's Hall, Philadelphia, October 4, 5, 6, 1876*. Washington, D.C.: Todd Brothers.

Association for the Advancement of Women. 1879. *Report of the Association for the Advancement of Women, 1878–1879*. Boston: Gunn, Bliss and Co.

———. 1880. *Annual Report of the Association for the Advancement of Women*. Dedham, Mass.: W. L. Wardle.

———. 1893. *Historical Account of the Association for the Advancement of Women, 1873–1893: Twenty-First Women's Congress, World's Columbian Exposition, Chicago, 1893*. Dedham, Mass.: Transcript Steam Job Print.

Athearn, Robert G. 1962. *Rebel of the Rockies: A History of the Denver and Rio Grande Western Railroad*. New Haven: Yale Univ. Press.

Babbitt, Mary King. 1912. *Maria Mitchell as Her Students Knew Her: An Address by Mary King Babbitt*. Poughkeepsie: The Enterprise Pub. Co.

Baily, Francis. 1836. "On a Remarkable Phenomenon That Occurs in Total and Annular Eclipses of the Sun." *Monthly Notices of the Royal Astronomical Society* 4 (2):15–19.

Bain, James S. 1940. *A Bookseller Looks Back: The Story of the Bains*. London: Macmillan.

Bairoch, Paul. 1982. "International Industrialization Levels from 1750 to 1980." *Journal of European Economic History* 11 (2):269-333.

Baldwin, Neil. 1995. *Edison: Inventing the Century*. New York: Hyperion.

Banning, Evelyn I. 1973. *Helen Hunt Jackson*. New York: Vanguard.

Barbour, Nelson H. 1871. *Evidences for the Coming of the Lord in 1873: or The Midnight Cry*. Rochester, N.Y.: Nelson H. Barbour.

Barker, George F. 1878. "On the Total Solar Eclipse of July 29th, 1878." *Proceedings of the American Philosophical Society* 18 (102):103–14.

———. 1879. "On the Results of the Spectroscopic Observation of the Solar Eclipse of July 29th, 1878." *American Journal of Science and Arts* [Third Series] 17 (98):121–25.

———. 1895. "Memoir of Henry Draper, 1837–1882." In *Biographical Memoirs*, vol. 3, pp. 81–139. Washington, D.C.: National Academy of Sciences.

Barnard, Harry. 1954. *Rutherford B. Hayes and His America*. Indianapolis: Bobbs-Merrill.

Barnes, J. K., et al. 1870. *Disasters on the Lakes [To Accompany Bill H. R. No. 602]: Letter Addressed to the Hon. Halbert E. Paine Relative to Storm Telegraphy. January 26, 1870*. 41st Congress, 2d Session, House of Representatives, Ex. Doc. No. 10 pt. 2. Washington, D.C.: Government Printing Office.

Barr, Alwyn. 1996. *Black Texans: A History of African Americans in Texas, 1528–1995*. 2nd ed. Norman: Univ. of Oklahoma Press.

Barr, E. Scott. 1963. "The Infrared Pioneers—III. Samuel Pierpont Langley." *Infrared Physics* 3 (4):195–206.

Barratt, Carrie Rebora. 2000. *Queen Victoria and Thomas Sully*. Princeton: Princeton Univ. Press.

Bartky, Ian R. 2000. *Selling the True Time: Nineteenth-Century Timekeeping in America*. Stanford: Stanford Univ. Press.

Bartlett, I. S., ed. 1918. *History of Wyoming*. Vol. 2. Chicago: S. J. Clarke.

Baum, Richard, and William Sheehan. 1997. *In Search of Planet Vulcan: The Ghost in Newton's Clockwork Universe*. New York: Plenum.

Baxter, James Phinney, ed. 1914. *Documentary History of the State of Maine*. Vol. 18. Portland: Maine Historical Society / Lefavor-Tower Company.

Bazerman, Charles. 1999. *The Languages of Edison's Light*. Cambridge, Mass.: MIT Press.

Beers, Dorothy Gondos. 1982. "The Centennial City, 1865–1876." In *Philadelphia: A 300-Year History*, edited by Russel F. Weigley, 417–70. New York: W. W. Norton.

Bell, Trudy E., and Robert B. Ariail. 2006. "Early Clark V: Maria Mitchell's 1872 Notes on Alvan Clark and Telescope Making." *Journal of the Antique Telescope Society* (27–28):34–39.

Bemis, Samuel Flagg. 1956. *John Quincy Adams and the Union*. New York: Knopf.

Bergland, Renée L. 2008. *Maria Mitchell and the Sexing of Science: An Astronomer among the American Romantics*. Boston: Beacon Press.

Bishop, William H. 1878. "A Night with Edison." *Scribner's Monthly* 17 (1):88–99.

The Black-Day, or, A Prospect of Doomsday. Exemplified in the Great and Terrible Eclipse, Which Will Happen on Friday the 22d of April, 1715. 1715. London: J. Read, R. Burleigh.

Blatch, Harriot Stanton, and Alma Lutz. 1940. *Challenging Years: The Memoirs of Harriot Stanton Blatch*. New York: G. P. Putnam's Sons.

Blavatsky, H. P. 1877. *Isis Unveiled: A Master-Key to the Mysteries of Ancient and Modern Science and Theology*. Vol. 1: *Science*. New York: J. W. Bouton.

Bliss, Sylvester. 1853. *Memoirs of William Miller, Generally Known as a Lecturer on the Prophecies, and the Second Coming of Christ*. Boston: Joshua V. Himes.

Block, Viola. 1970. *History of Johnson County and Surrounding Areas*. Waco: Texian Press.

Booker, Margaret Moore. 2007. *Among the Stars: The Life of Maria Mitchell; Astronomer, Educator, Women's Rights Activist*. Nantucket, Mass.: Mill Hill Press.

Bowles, Samuel. 1869. *Our New West: Records of Travel between the Mississippi River and the Pacific Ocean*. Hartford: Hartford Publishing.

Brewster, David. 1855. *Memoirs of the Life, Writings, and Discoveries of Sir Isaac Newton*. Vol. 2. Edinburgh: Thomas Constable.

Broughton, Peter. 1996. "James Craig Watson (1838–1880)." *Journal of the Royal Astronomical Society of Canada* 90 (2):74–81.

Brown, Dee. 1966. *The Year of the Century: 1876*. New York: Charles Scribner's Sons.

Brown, J. Willard. 1896. *The Signal Corps, U. S. A. in the War of the Rebellion*. Boston: U. S. Veteran Signal Corps Association.

Brown, Joseph G. 1898. *The History of Equal Suffrage in Colorado, 1868–1898*. Denver: News Job Printing Co.

Brown, Larry. 1995. "Lillian Heath Nelson, Pioneer Woman Doctor." *Wyoming History Journal* 67 (2): 46–47.

Bruce, Robert V. 1973. *Bell: Alexander Graham Bell and the Conquest of Solitude*. Boston: Little, Brown.

———. 1987. *The Launching of Modern American Science, 1846–1876*. New York: Knopf.

Bryan, George S. 1926. *Edison: The Man and His Work*. Garden City, N.Y.: Garden City Publishing.

Butler, Jonathan M. 1991. *Softly and Tenderly Jesus Is Calling: Heaven and Hell in American Revivalism, 1870–1920*. Brooklyn: Carlson.

Byrd, A. J. 1879. *History and Description of Johnson County and Its Principal Towns. . . .* Marshall, Tex.: Jennings Bros.

Campbell, W. W. 1909. "The Closing of a Famous Astronomical Problem." *Popular Science Monthly* 74 (May):494–503.

———. 1924. "Biographical Memoir: Simon Newcomb, 1835–1909." In *Memoirs of the National Academy of Sciences*, vol. 17, pp. 1–18. Washington, D.C.: Government Printing Office.

Cannon, Annie J. 1915a. "Mrs. Henry Draper." *Science* [New Series] 41 (1054):380–82.

———. 1915b. "The Henry Draper Memorial." *Journal of the Royal Astronomical Society of Canada* 9 (5):203–15.

Carter, Bill, and Merri Sue Carter. 2006. *Simon Newcomb: America's Unofficial Astronomer Royal*. St. Augustine, Fla.: Matanzas.

Case, Theo S. 1878. "The Solar Eclipse of July 29, 1878." *The Western Review of Science and Industry* 2 (5):263–69.

Chang, Jung. 2013. *Empress Dowager Cixi: The Concubine Who Launched Modern China*. New York: Knopf.

Chicago Astronomical Society. 1878. *The Solar Eclipse of July 29, 1878*. Chicago: Evening Journal Book and Job Printing House.

Clarke, Edward H. 1873. *Sex in Education; or, A Fair Chance for the Girls*. Boston: James R. Osgood.

Clarke, John Jackson. 1929. "Reminiscenses [sic] of Wyoming in the Seventies and Eighties." *Annals of Wyoming* 6 (1 and 2):225–36.

Cleminshaw, C. H. 1946. "The Founding of the Cincinnati Observatory." *Astronomical Society of the Pacific Leaflets* 5 (208):65–72.

Coddington, E. F. 1900. "Observations of the Watson Asteroids." *Astronomische Nachrichten* 153 (13):225–34.

Cohen, Richard. 2010. *Chasing the Sun: The Epic Story of the Star That Gives Us Life*. New York: Random House.

Comfort, George Fisk, and Anna Manning Comfort. 1874. *Woman's Education and Woman's Health: Chiefly in Reply to "Sex in Education."* Syracuse: Thos. W. Durston.

Comstock, George C. 1895. "Memoir of James Craig Watson, 1838–1880." In *Biographical Memoirs*, vol. 3, pp. 43–58. Washington, D.C.: National Academy of Sciences.

Comte, Auguste. 1835. *Cours de Philosophie Positive*. Vol. 2. Paris: Bachelier.

Congressional Record: Containing the Proceedings and Debates of the Forty-Fifth Congress, Second Session. Vol. 7, Part 3. 1878. Washington, D.C.: Government Printing Office.

Conot, Robert. 1979. *A Streak of Luck*. New York: Seaview.

Cook, Alan. 1998. *Edmond Halley: Charting the Heavens and the Seas*. Oxford: Clarendon.

Cooper, James Fenimore. 1869. "The Eclipse." *Putnam's* 4 (21):352–59.

Copernicus, Kepler [pseud.]. 1869. "The Eclipse Party." *Nassau Literary Magazine* 26 (2):104–12.

Corbett, Hoye and Co. 1878. *Corbett, Hoye & Co.'s Sixth Annual City Directory of the Inhabitants, Institutions, Incorporated Companies, Manufacturing Establishments, Business, Business Firms, etc., in the City of Denver, for 1878*. Denver: Corbett, Hoye and Co.

Corbin, Brenda G. 2007. "Etienne Leopold Trouvelot (1827–1895), the Artist and Astronomer." In *Library and Information Services in Astronomy V: Common Challenges, Uncommon Solutions; Proceedings of a Conference Co-hosted by the Harvard-Smithsonian Center for Astrophysics and Massachusetts Institute of Technology, Cambridge, Massachusetts, USA, 18–21 June 2006*, edited by Sandra Ricketts, Christina Birdie, and Eva Isaksson, 352–60. San Francisco: Astronomical Society of the Pacific.

Cortie, A. L. 1921. "Sir Norman Lockyer, 1836-1920." *Astrophysical Journal* 53 (4):233–48.

Cottam, Stella, et al. 2011. "The Total Solar Eclipses of 7 August 1869 and 29 July 1878 and the Popularisation of Astronomy in the USA as Reflected in the *New York Times*." In *Highlighting the History of Astronomy in the Asia-Pacific Region: Proceedings of the ICOA-6 Conference*, edited by Wayne Orchiston et al., 339–75. New York: Springer.

Cottam, Stella, and Wayne Orchiston. 2014. *Eclipses, Transits, and Comets of the Nineteenth Century: How America's Perception of the Skies Changed*. Springer International.

Cox, John D. 2002. *Storm Watchers: The Turbulent History of Weather Prediction from Franklin's Kite to El Niño*. Hoboken: John Wiley and Sons.

Craig, Newton N. (Nute). 1931. *Thrills, 1861 to 1887*. Oakland: N. N. Craig.

Cranston, Sylvia. 1993. *HPB: The Extraordinary Life and Influence of Helena Blavatsky, Founder of the Modern Theosophical Movement*. New York: G. P. Putnam's Sons.

Creese, Mary R. S. 1998. *Ladies in the Laboratory? American and British Women in Science, 1800–1900*. Lanham, Md.: Scarecrow Press.

Crelinsten, Jeffrey. 2006. *Einstein's Jury: The Race to Test Relativity*. Princeton: Princeton Univ. Press.

Crofutt, George A. 1878. *Crofutt's New Overland Tourist and Pacific Coast Guide, Containing a Condensed and Authentic Description of over One Thousand Two Hundred Cities, Towns, Villages, Stations, Government Fort and Camps, Mountains, Lakes, Rivers, Sulphur, Soda and Hot Springs, Scenery, Watering Places, and Summer Resorts. . . .* Chicago: Overland Publishing Co.

Curtis, Heber D. 1938. "James Craig Watson, 1838–1880." *Michigan Alumnus Quarterly Review* 44 (24):306–13.

Dall, Caroline H. 1881. *My First Holiday; or, Letters Home from Colorado, Utah, and California.* Boston: Roberts Brothers.

Daniels, George H. 1967. "The Pure-Science Ideal and Democratic Culture." *Science* [New Series] 156 (3783):1699–1705.

Dawson, Thomas F., and F. J. V. Skiff. (1879) 1980. *The Ute War: A History of the White River Massacre.* Denver: Tribune Publishing House. Reprint with introduction by Michael McNierney, Boulder: Johnson.

De La Rue, Warren. 1862. *The Bakerian Lecture on the Total Solar Eclipse of July 18th, 1860, Observed at Rivabellosa, near Miranda de Ebro, in Spain.* London: Taylor and Francis.

Decker, Peter R. 2004. *The Utes Must Go! American Expansion and the Removal of a People.* Golden, Colo.: Fulcrum.

Degni, Rev. J. M. 1878. "The Total Solar Eclipse of July 29th, 1878." *American Catholic Quarterly Review* 3:635–48.

DeGraaf, Leonard. 2013. *Edison and the Rise of Innovation.* New York: Sterling.

Dick, Steven J. 2003. *Sky and Ocean Joined: The U.S. Naval Observatory, 1830–2000.* Cambridge, U.K.: Cambridge Univ. Press.

Dick, Steven J., Wayne Orchiston, and Tom Love. 1998. "Simon Newcomb, William Harkness and the Nineteenth-Century American Transit of Venus Expeditions." *Journal for the History of Astronomy* 29 (3):221–55.

Dickson, W. K. L., and Antonia Dickson. 1894. *The Life and Inventions of Thomas Alva Edison.* New York: Thomas Y. Crowell.

Dorsett, Lyle W. 1977. *The Queen City: A History of Denver.* Boulder: Pruett.

Drake, Benjamin. 1858. *Life of Tecumseh, and of His Brother the Prophet; with a Historical Sketch of the Shawanoe Indians.* Cincinnati: Anderson, Gates and Wright.

Draper, Henry. 1878a. "The Solar Eclipse of July 29, 1878." *Science News* 1 (2):17–19.

———. 1878b. "The Solar Eclipse of July 29th, 1878." *American Journal of Science and Arts* [Third Series] 16 (93):227–30.

———. 1878c. "The Total Solar Eclipse of July 29th, 1878." *Journal of the Franklin Institute* 106 (4):217–20.

DuBois, Ellen Carol, ed. 1992. *The Elizabeth Cady Stanton–Susan B. Anthony Reader.* Rev. ed. Boston: Northeastern Univ. Press.

Duffey, Mrs. E. B. 1874. *No Sex in Education; or, An Equal Chance for Both Girls and Boys.* Philadelphia: J. M. Stoddart.

Dunraven, Earl of [Windham Thomas Wyndham-Quin]. 1876. *The Great Divide: Travels in the Upper Yellowstone in the Summer of 1874.* London: Chatto and Windus.

Dupree, A. Hunter. 1957. *Science in the Federal Government: A History of Policies and Activities to 1940.* Cambridge, Mass.: Belknap Press of Harvard Univ. Press.

Dyer, Frank Lewis, and Thomas Commerford Martin. 1910. *Edison: His Life and Inventions.* 2 vols. New York: Harper and Bros.

Dyson, F. W., A. S. Eddington, and C. Davidson. 1920. "A Determination of the Deflection of Light by the Sun's Gravitational Field, from Observations Made at the Total Eclipse of May 29, 1919." *Philosophical Transactions of the Royal Society of London. Series A, Containing Papers of a Mathematical or Physical Character* 220 (571–581): 291–333.

Eddy, John A. 1972. "Thomas A. Edison and Infra-red Astronomy." *Journal for the History of Astronomy* 3 (3):165–87.

———. 1973. "The Great Eclipse of 1878." *Sky and Telescope* 45 (6):340–46.

———. 1979. "Edison the Scientist." *Applied Optics* 18 (22):3737–50.

Edison, Thomas A. 1879. "On the Use of the Tasimeter for Measuring the Heat of the Stars and of the Sun's Corona." *American Journal of Science and Arts* [Third Series] 17 (97):52–54.

———. 1904. "The Beginnings of the Incandescent Lamp." *Electrical World and Engineer* 43 (10):431–32.

———, et al. 1989. *The Papers of Thomas A. Edison.* Vol. 1: *The Making of an Inventor, February 1847–June 1873.* Edited by Reese V. Jenkins, Leonard S. Reich, Paul B. Israel, Toby Appel, Andrew J. Butrica, Robert A. Rosenberg, Keith A. Nier, Melodie Andrews, and Thomas E. Jeffrey. Baltimore: Johns Hopkins Univ. Press.

———, et al. 1991. *The Papers of Thomas A. Edison.* Vol. 2: *From Workshop to Laboratory, June 1873–March 1876.* Edited by Robert A. Rosenberg, Paul B. Israel, Keith A. Nier, and Melodie Andrews. Baltimore: Johns Hopkins Univ. Press.

———, et al. 1994. *The Papers of Thomas A. Edison.* Vol. 3: *Menlo Park: The Early Years, April 1876–December 1877.* Edited by Robert A. Rosenberg, Paul B. Israel, Keith A. Nier, and Martha J. King. Baltimore: Johns Hopkins Univ. Press.

———, et al. 1998. *The Papers of Thomas A. Edison.* Vol. 4: *The Wizard of Menlo Park, 1878.* Edited by Paul B. Israel, Keith A. Nier, and Louis Carlat. Baltimore: Johns Hopkins Univ. Press.

———, et al. 2004. *The Papers of Thomas A. Edison.* Vol. 5: *Research to Development at Menlo Park, January 1879–March 1881.* Edited by Paul B. Israel, Louis Carlat, David Hochfelder, and Keith A. Nier. Baltimore: Johns Hopkins Univ. Press.

Edmunds, R. David. 1983. *The Shawnee Prophet.* Lincoln: Univ. of Nebraska Press.

Elliott, Clark A. 1996. *History of Science in the United States: A Chronology and Research Guide.* New York: Garland.

Ellis, Constance Dimock, ed. 1961. *The Magnificent Enterprise: A Chronicle of Vassar College.* Compiled by Dorothy A. Plum and George B. Dowell, Vassar College Centennial Committee. Poughkeepsie: Vassar College.

Emmitt, Robert. (1954) 2000. *The Last War Trail: The Utes and the Settlement of Colorado, with a New Introduction by Andrew Gulliford and Afterword by Charles Wilkinson.* Norman: Univ. of Oklahoma Press. Reprint, Boulder: Univ. Press of Colorado.

Fenno, Frank H. 1878. *The Science and Art of Elocution; or, How to Read and Speak.* Philadelphia: John E. Potter.

Field, Cynthia R., Richard E. Stamm, and Heather P. Ewing. 1993. *The Castle: An Illustrated History of the Smithsonian Building.* Washington, D.C.: Smithsonian Institution Press.

Fishell, Dave. 1999. *Towers of Healing: The First 125 Years of Denver's Saint Joseph Hospital.* Denver: Saint Joseph Hospital Foundation.

Fontenrose, Robert. 1973. "In Search of Vulcan." *Journal for the History of Astronomy* 4 (3):145–58.

Forbes, James D. 1843. "Sur l'Éclipse Totale de Soleil du 8 Juillet 1842." *Bibliothèque Universelle de Genève* 48 (December):361–68.

Fossett, Frank. 1880. *Colorado, Its Gold and Silver Mines, Farms and Stock Ranges, and Health and Pleasure Resorts. Tourist's Guide to the Rocky Mountains.* 2nd ed. New York: C. G. Crawford.

Fox, Edwin M. 1879. "Edison's Inventions. II. The Carbon Button and Its Offspring." *Scribner's Monthly* 18 (3):446–55.

Freeberg, Ernest. 2013. *The Age of Edison: Electric Light and the Invention of Modern America.* New York: Penguin.

Freeman, James H. 1937. "A Total Eclipse of the Sun, July 29, 1878." *Popular Astronomy* 45 (9):491–92.

Friedel, Robert, and Paul Israel with Bernard S. Finn. 2010. *Edison's Electric Light: The Art of Invention.* Baltimore: Johns Hopkins Univ. Press.

Friend, John C. 1943. "Early History of Carbon County." *Annals of Wyoming* 15 (3):280–96.

Frieze, Henry S., et al. 1882. *The Memorial Addresses Delivered in University Hall, November 26, 1880, at the Funeral of Professor James Craig Watson, Ph.D, LL.D., Professor in the University from 1859 to 1879.* Ann Arbor: University of Michigan.

Frost, Edwin B. 1910. "Biographical Memoir of Charles Augustus Young 1834–1908." In *Biographical Memoirs*, vol. 7, pp. 91–114. Washington, D.C.: National Academy of Sciences.

Frye, Elnora L. 1990. *Atlas of Wyoming Outlaws at the Territorial Penitentiary.* Laramie: Jelm Mountain Publications.

Galbraith, R. M. 1922. "With Edison on Union Pacific When Incandescent Light Was Invented." *Union Pacific Magazine* 1 (9):4–5, 25.

Gaustad, Edwin S, ed. 1974. *The Rise of Adventism: Religion and Society in Mid-Nineteenth-Century America.* New York: Harper and Row.

Giberti, Bruno. 2002. *Designing the Centennial: A History of the 1876 International Exhibition in Philadelphia.* Lexington: Univ. of Kentucky Press.

Gilmore, Charles M., ed. 1876. *The Herald Guide Book and Directory to the Centennial Exposition. . . .* Philadelphia: Charles M. Gilmore.

Gingerich, Owen. 1981. "Astronomical Scrapbook: Eighteenth-Century Eclipse Paths." *Sky and Telescope* 62 (4):324–27.

Glassford, W. A. 1891. "Historical Sketch of the Signal Corps, U.S. Army." *Journal of the Military Service Institution of the United States* 12 (54):1325–38.

Goetzmann, William H. 1966. *Exploration and Empire: The Explorer and the Scientist in the Winning of the American West.* New York: Alfred A. Knopf.

Goldfield, David. 2011. *America Aflame: How the Civil War Created a Nation.* New York: Bloomsbury.

Goode, George Brown, ed. 1897. *The Smithsonian Institution 1846–1896.* Washington, D.C.: City of Washington.

Gordon, Ann D., ed. 1997. *The Selected Papers of Elizabeth Cady Stanton and Susan B. Anthony.* Vol. 3: *National Protection for National Citizens, 1873 to 1880.* New Brunswick, N.J.: Rutgers Univ. Press.

Greely, A. W. 1916. "Cleveland Abbe." *Science* [New Series] 44 (1142):703–4.

Grier, David Alan. 2005. *When Computers Were Human.* Princeton: Princeton Univ. Press.

Guy, Joseph, and Thomas Keith. 1845. *Guy's Elements of Astronomy, and an Abridgment of Keith's New Treatise on the Use of the Globes.* 30th ed. Philadelphia: Thomas, Cowperthwait.

Hackett, P. H., and R. C. Roach. 2004. "High Altitude Cerebral Edema." *High Altitude Medicine and Biology* 5 (2):136–46.

Hale, George E. 1895. "Arthur Cowper Ranyard." *Astrophysical Journal* 1 (2):168–69.

Hall, Frank. 1895. *History of the State of Colorado.* Vol. 4. Chicago: Blakely Printing Co.

Hall, Isaac H., and Oren Root. 1890. "Dr. Christian Henry Frederick Peters, Born September 19, 1813, Died July 18, 1890." *Hamilton Literary Monthly* 25 (3):iii–xxxi.

Hall, Roger A. 2001. *Performing the American Frontier, 1870–1906*. Cambridge, U.K.: Cambridge Univ. Press.

Halley, Edmund. 1715. "Observations of the Late Total Eclipse of the Sun on the 22[d] of April Last Past . . ." *Philosophical Transactions* 29 (343):245–62.

Hansen, Julie M. Vinter. 1943. "Positions of the Watson Asteroids: 101 Helena, 103 Hera, 119 Althaea, 128 Nemesis, 161 Athor and 179 Klytaemnestra." *Lick Observatory Bulletin* 514:161–62.

Harper, Ida Husted. 1899. *The Life and Work of Susan B. Anthony: Including Public Addresses, Her Own Letters and Many from Her Contemporaries during Fifty Years*. 2 vols. Indianapolis: Bowen-Merrill.

Hearnshaw, John. 2010. "Auguste Comte's Blunder: An Account of the First Century of Stellar Spectroscopy and How It Took One Hundred Years to Prove That Comte Was Wrong!" *Journal of Astronomical History and Heritage* 13 (2):90–104.

Heaton, John W. 2005. *The Shoshone-Bannocks: Culture and Commerce at Fort Hall, 1870–1940*. Lawrence: Univ. Press of Kansas.

Herodotus. (1920) 1990. *Herodotus*. Translated by A. D. Godley. Vol. 1. London: William Heinemann; New York: G. P. Putnam's Sons. Reprint, Cambridge, Mass.: Harvard Univ. Press.

Hetherington, Norriss S. 1975. "Financing Education and Science in Nineteenth-Century America: The Case of Cleveland Abbe, the Chicago Astronomical Society and the First University of Chicago." *Journal of the Illinois State Historical Society* 68 (4):319–23.

———. 1976. "Cleveland Abbe and a View of Science in Mid-Nineteenth-Century America." *Annals of Science* 33 (1):31–49.

Hirshfeld, Alan. 2006. *The Electric Life of Michael Faraday*. New York: Walker.

Historical and Philosophical Society of Ohio. 1944. *The Centenary of the Cincinnati Observatory, 1843–1943*. Cincinnati: The Historical and Philosophical Society of Ohio and Univ. of Ohio.

Hodgson, Richard, et al. 1885. "Report of the Committee Appointed to Investigate Phenomena Connected with the Theosophical Society." *Proceedings of the Society for Psychical Research* 3 (9):201–400.

Hoffleit, Dorrit. 1983. *Maria Mitchell's Famous Students; and, Comets over Nantucket*. Cambridge, Mass.: American Association of Variable Star Observers.

Holden, Edward S. 1881. "Report of the Director of the Observatory and Professor of Astronomy to the Board of Regents." In *Annual Report of the Board of Regents of the University of Wisconsin for the Fiscal Year Ending September 30, 1881*, 29–32. Madison: David Atwood, State Printer.

———. 1882. "Report of the Director of the Observatory and Professor of Astronomy to the Board of Regents." In *Annual Report of the Board of Regents of the University of Wisconsin for the Fiscal Year Ending September 30, 1882*, 31–37. Madison: David Atwood, State Printer.

———. 1896. "Notices from the Lick Observatory." *Publications of the Astronomical Society of the Pacific* 8 (46):27–31.

Hoskin, Michael. 2011. *Discoverers of the Universe: William and Caroline Herschel*. Princeton: Princeton Univ. Press.

Hounshell, David A. 1980. "Edison and the Pure Science Ideal in 19th-Century America." *Science* [New Series] 207 (4431):612–17.

Howe, Frederic C. 1925. *The Confessions of a Reformer*. New York: Charles Scribner's Sons.

Howe, Julia Ward, ed. 1874. *Sex* and *Education: A Reply to Dr. E. H. Clarke's "Sex in Education."* Boston: Roberts Brothers.

———. 1884. "Maria Mitchell." In *Our Famous Women: An Authorized Record of Their Lives and Deeds*, edited by the publisher, 437–61. Hartford: A. D. Worthington.

———. 1900. *Reminiscences 1819–1899*. Boston: Houghton Mifflin.

———. 1906. "What Life Means to Me." *Cosmopolitan* 41 (3):285–89.

Howells, W. D. 1876. "A Sennight of the Centennial." *Atlantic Monthly* 38 (125):92–107.

Howes, Annie G., et al. 1885. *Health Statistics of Women College Graduates. Report of a Special Committee of the Association of Collegiate Alumnae, Annie G. Howes, Chairman, Together with Statistical Tables Collated by the Massachusetts Bureau of Statistics of Labor.* Boston: Wright and Potter, State Printers.

Howlett, Rev. W. J. (1908) 1954. *Life of the Right Reverend Joseph P. Machebeuf, D. D.* Pueblo: Franklin Press. Reprint, Denver: Register College of Journalism.

Hughes, David W. 1983. "On Seeing Stars (Especially up Chimneys)." *Quarterly Journal of the Royal Astronomical Society* 24 (3):246–57.

Hughes, Ivor, and David Ellis Evans. 2011. *Before We Went Wireless: David Edward Hughes, FRS; His Life, Inventions and Discoveries (1829–1900)*. Bennington, Vt.: Images from the Past.

Humphreys, William J. 1919. "Biographical Memoir of Cleveland Abbe 1838–1916." In *Biographical Memoirs*, vol. 8, pp. 469–508. Washington, D.C.: National Academy of Sciences.

Husband, Joseph. 1917. *The Story of the Pullman Car.* Chicago: A. C. McClurg.

Isaacson, Walter. 2003. *Benjamin Franklin: An American Life.* New York: Simon and Schuster.

Israel, Paul. 1998. *Edison: A Life of Invention.* New York: John Wiley and Sons.

Jacobi, Mary Putnam. 1877. *The Question of Rest for Women During Menstruation.* New York: G. P. Putnam's Sons.

Jaffe, Mark. 2000. *The Gilded Dinosaur: The Fossil War Between E. D. Cope and O. C. Marsh and the Rise of American Science.* New York: Crown.

Jehl, Francis. 1937. *Menlo Park Reminiscences.* Vol. 1. Dearborn: Edison Institute.

Jensen, Billie Barnes. 1973. "Colorado Woman Suffrage Campaigns of the 1870s." *Journal of the West* 12 (2):254–71.

Jensen, J. Vernon. 1993. "Thomas Henry Huxley's Address at the Opening of the Johns Hopkins University in September 1876." *Notes and Records of the Royal Society of London* 47 (2):257–69.

Jones, Bence. 1870. *The Life and Letters of Faraday.* Vol. 1. London: Longmans, Green.

Jones, Bessie Zaban, and Lyle Gifford Boyd. 1971. *The Harvard College Observatory.* Cambridge, Mass.: Belknap Press of Harvard Univ. Press.

Jones, Francis Arthur. 1908. *Thomas Alva Edison: Sixty Years of an Inventor's Life.* New York: Thomas Y. Crowell.

Jones, Payson. 1940. *A Power History of the Consolidated Edison System, 1878–1900.* New York: Consolidated Edison Co. of New York.

Jones, William C., and Kenton Forrest. 1973. *Denver: A Pictorial History from Frontier Camp to Queen City of the Plains.* Boulder: Pruett.

Jonnes, Jill. 2003. *Empires of Light: Edison, Tesla, Westinghouse, and the Race to Electrify the World.* New York: Random House.

Josephson, Matthew. 1959. *Edison: A Biography.* New York: McGraw-Hill.

Kasson, Joy S. 2000. *Buffalo Bill's Wild West: Celebrity, Memory, and Popular History.* New York: Hill and Wang.

Kendall, Phebe Mitchell, ed. 1896. *Maria Mitchell: Life, Letters, and Journals.* Boston: Lee and Shepard.

Kennedy, J. E., and S. D. Hanson. 1996. "Excerpts from Simon Newcomb's Diary of 1860." *Journal of the Royal Astronomical Society of Canada* 90 (5/6):292–303.

Kennelly, Arthur E. 1933. "Biographical Memoir of Thomas Alva Edison, 1847–1931." In *Biographical Memoirs*, vol. 15, pp. 287–304. Washington, D.C.: National Academy of Sciences.

Kevles, Daniel J. 1995. *The Physicists: The History of a Scientific Community in Modern America.* Rev. ed. Cambridge, Mass.: Harvard Univ. Press.

———. 2013. "Not a Hundred Millionaires: The National Academy and the Expansion of Federal Science in the Gilded Age." *Issues in Science and Technology* 29 (2):37–46.

Kinnaman, Daniel L. 2013. *Rawlins, Wyoming: The Territorial Years, 1868–1890.* Rawlins: Kinnaman Supply Co.

Kirkland, A. H. 1906. *First Annual Report of the Superintendent for Suppressing the Gypsy and Brown-Tail Moths.* Boston: Wright and Potter Printing Co., State Printers.

Kirkwood, Daniel. 1888. *The Asteroids, or Minor Planets Between Mars and Jupiter.* Philadelphia: J. B. Lippincott.

Koenigsberg, Allen. 1987. "The First 'Hello!' Thomas Edison, the Phonograph and the Telephone." *Antique Phonograph Monthly* 8 (6):3–9.

Kohlstedt, Sally Gregory. 1978a. "In from the Periphery: American Women in Science, 1830–1880." *Signs* 4 (1):81–96.

———. 1978b. "Maria Mitchell: The Advancement of Women in Science." *New England Quarterly* 51 (1):39–63.

Kyle, Richard. 1998. *The Last Days Are Here Again: A History of the End Times.* Grand Rapids: Baker.

Labaree, Leonard W., ed. 1961. *The Papers of Benjamin Franklin.* Vol. 3: *January 1, 1745, through June 30, 1750.* New Haven: Yale Univ. Press.

Langley, Samuel Pierpont. 1889. "The History of a Doctrine." *American Journal of Science and Arts* [Third Series] 37 (217):1–23.

———. 1891. *The New Astronomy.* Boston: Houghton Mifflin.

Lankford, John, with Ricky L. Slavings. 1997. *American Astronomy: Community, Careers, and Power, 1859–1940.* Chicago: Univ. of Chicago Press.

Launay, Françoise. 2012. *The Astronomer Jules Janssen: A Globetrotter of Celestial Physics.* Translated by Storm Dunlop. New York: Springer.

Lehmann, Rudolph Chambers. 1897. *In Cambridge Courts.* London: H. Henry.

Leng, John. 1877. *America in 1876: Pencillings during a Tour in the Centennial Year; with a Chapter on the Aspects of American Life.* Dundee: *Dundee Advertiser* Office.

Leonard, Stephen J., and Thomas J. Noel. 1990. *Denver: Mining Camp to Metropolis.* Niwot, Colo.: Univ. Press of Colorado.

Lequeux, James. 2013. *Le Verrier—Magnificent and Detestable Astronomer.* Edited by William Sheehan. Translated by Bernard Sheehan. New York: Springer.

Leslie, Mrs. Frank. 1877. *California: A Pleasure Trip from Gotham to the Golden Gate (April, May, June, 1877).* New York: G. W. Carleton.

Levenson, Thomas. 2015. *The Hunt for Vulcan: . . . And How Albert Einstein Destroyed a Planet, Discovered Relativity, and Deciphered the Universe.* New York: Random House.

Leyendecker, Liston Edgington. 1992. *Palace Car Prince: A Biography of George Mortimer Pullman.* Niwot, Colo.: Univ. Press of Colorado.

Lightman, Alan, and Roberta Brawer. 1990. *Origins: The Lives and Worlds of Modern Cosmologists.* Cambridge, Mass.: Harvard Univ. Press.

Lindley, David. 2004. *Degrees Kelvin: A Tale of Genius, Invention, and Tragedy*. Washington, D.C.: Joseph Henry Press.

Littmann, Mark, Fred Espenak, and Ken Willcox. 2008. *Totality: Eclipses of the Sun*. 3rd ed. Oxford: Oxford Univ. Press.

Lockyer, J. Norman. 1874. *Contributions to Solar Physics*. London: Macmillan.

——. 1878a. "The Coming Total Solar Eclipse." *Nature* 17 (442):481–83.

——. 1878b. "The Coming Total Solar Eclipse II." *Nature* 17 (443):501–3.

——. 1878c. "What the Sun Is Made of." *Nineteenth Century* 4 (17):75–87.

——. 1896. "The Story of Helium." *Nature* 53 (1371–72):319–22, 342–46.

Lockyer, T. Mary and Winifred L. Lockyer, with H. Dingle. 1928. *Life and Work of Sir Norman Lockyer*. London: Macmillan.

Logan, Herschel C. 1954. *Buckskin and Satin: The Life of Texas Jack (J. B. Omohundro) Buckskin Clad Scout, Indian Fighter, Plainsman, Cowboy, Hunter, Guide and Actor, and His Wife Mlle. Morlacci Premiere Danseuse in Satin Slippers*. Harrisburg: Stackpole.

Lu, Lingfeng, and Huifang Li. 2013. "Chinese Records of the 1874 Transit of Venus." *Journal of Astronomical History and Heritage* 16 (1):45–54.

Lynn, W. T. 1892. "Copernicus and Mercury." *Observatory* 15 (191):321–22.

Macfarlane, Alexander. 1916. *Lectures on Ten British Mathematicians of the Nineteenth Century*. Mathematical Monographs 17. New York: John Wiley and Sons.

Machebeuf, Rt. Rev. Joseph Projectus. 1877. *Woman's Suffrage. A Lecture, Delivered in the Catholic Church of Denver, Colorado*. Denver: Tribune Steam Printing House.

Mack, Pamela E. 1990. "Strategies and Compromises: Women in Astronomy at Harvard College Observatory, 1870–1920." *Journal for the History of Astronomy* 21 (1):65–76.

Maury, T. B. 1871. "The Telegraph and the Storm: The United States Signal Service." *Harper's New Monthly Magazine* 43 (255):398–418.

Mazzotti, Massimo. 2010. "The Jesuit on the Roof: Observatory Sciences, Metaphysics, and Nation-Building." In *The Heavens on Earth: Observatories and Astronomy in Nineteenth-Century Science and Culture*, edited by David Aubin, Charlotte Bigg, and H. Otto Sibum, 58–85. Durham: Duke Univ. Press.

McClure, A. K. 1905. *Old Time Notes of Pennsylvania*. Vol. 2. Philadelphia: John C. Winston.

McClure, J. B., ed. 1889. *Edison and His Inventions, Including the Many Incidents, Anecdotes, and Interesting Particulars Connected with the Early and Late Life of the Great Inventor . . .* Chicago: Rhodes and McClure.

McKivigan, John. 2008. *Forgotten Firebrand: James Redpath and the Making of Nineteenth-Century America*. Ithaca: Cornell Univ. Press.

Mead, Rebecca J. 2004. *How the Vote Was Won: Woman Suffrage in the Western United States, 1848–1914*. New York: New York Univ. Press.

Meadows, A. J. 2008. *Science and Controversy: A Biography of Sir Norman Lockyer, Founder Editor of Nature*. 2nd ed. London: Macmillan.

Meeker, Josephine. (1879) 1976. *The Ute Massacre! Brave Miss Meeker's Captivity! Her Own Account of It*. Philadelphia: The Old Franklin Publishing House. Reprint, New York: Garland Publishing.

Menand, Louis. 2001. *The Metaphysical Club: A Story of Ideas in America*. New York: Farrar, Straus, and Giroux.

Mendenhall, T. C. 1882. "On the Influence of Time on the Change in the Resistance of the Carbon Disk of Edison's Tasimeter." *American Journal of Science and Arts* [Third Series] 24 (139):43–46.

Merrick, Sara Newcomb. 1910. "John and Simon Newcomb: The Story of a Father and Son." *McClure's* 35 (6):677–87.

Millard, Andre, Duncan Hay, and Mary Grassick. 1995. *Historic Furnishings Report, Edison Laboratory: Edison National Historic Site, West Orange, New Jersey.* Harpers Ferry: Division of Historic Furnishings, Harpers Ferry Center, National Park Service.

Miller, Howard S. 1970. *Dollars for Research: Science and Its Patrons in Nineteenth-Century America.* Seattle: Univ. of Washington Press.

Miller, Mark E. 1997. *Hollow Victory: The White River Expedition of 1879 and the Battle of Milk Creek.* Niwot, Colo.: Univ. Press of Colorado.

Mitchel, F. A. 1887. *Ormsby MacKnight Mitchel: Astronomer and General.* Boston: Houghton Mifflin.

Mitchell, Henry. 1890. "Maria Mitchell." *Proceedings of the American Academy of Arts and Sciences* 25:331–43.

Mitchell, Maria. 1869. "The Total Eclipse of 1869." *Hours at Home* 9 (6):555–60.

———. 1889. "Maria Mitchell's Reminiscences of the Herschels." *Century Illustrated Magazine*, 38 (6):903–9.

Moody, Dwight Lyman. 1877. *New Sermons, Addresses, and Prayers.* Cincinnati: Henry S. Goodspeed and Co.

Moody, Marshall D. 1953. "The Meeker Massacre." *Colorado Magazine* 30 (2):91–104.

Moore, Peter. 2015. *The Weather Experiment: The Pioneers Who Sought to See the Future.* New York: Farrar, Straus and Giroux.

Morris, Maud Burr. 1918. "An Old Washington Mansion (2017 I Street Northwest)." *Records of the Columbia Historical Society* 21:114–28.

Morris, Roy Jr. 2003. *Fraud of the Century: Rutherford B. Hayes, Samuel Tilden, and the Stolen Election of 1876.* New York: Simon and Schuster.

Morton, Henry. 1889. "Electricity in Lighting." *Scribner's* 6 (2):176–200.

Moyer, Albert E. 1992. *A Scientist's Voice in American Culture: Simon Newcomb and the Rhetoric of Scientific Method.* Berkeley: Univ. of California Press.

Nath, Biman B. 2013. *The Story of Helium and the Birth of Astrophysics.* New York: Springer.

National Academy of Sciences. 1909. *Report of the National Academy of Sciences for the Year 1908.* Washington, D.C.: Government Printing Office.

National Encyclopaedia, The: A Dictionary of Universal Knowledge. Vol. 13. 1867. London: William Mackenzie.

Nautical Almanac Office. 1877. *Almanac for the Use of Navigators, from the American Ephemeris and Nautical Almanac, for the Year 1878.* 2nd ed. Washington, D.C.: Bureau of Navigation.

———. 1878. *The Total Eclipse of the Sun on July 29, 1878. A Supplement to the American Ephemeris and Nautical Almanac.* Washington, D.C.: Nautical Almanac Office.

Nevins, Allan. 1927. *The Emergence of Modern America, 1865–1878.* New York: Macmillan.

Newcomb, Simon. 1860. "On the Supposed Intra-Mercurial Planets." *Astronomical Journal* 6 (141):162–63.

———. 1869. "A Proposed Arrangement for Observing the Corona, and Searching for Intra-Mercurial Planets during a Total Eclipse of the Sun." *American Journal of Science and Arts* [Second Series] 47 (141):413–15.

———. 1876. "Abstract Science in America, 1776–1876." *North American Review* 122 (250):88–123.

———. 1878a. "Eclipses of the Sun." *Princeton Review* 2 (November):848–64.

————. 1878*b*. *Popular Astronomy*. New York: Harper and Brothers.

————, ed. 1880. *Observations of the Transit of Venus, December 8–9, 1874. Made and Reduced under the Direction of the Commission Created by Congress*. Washington, D.C.: Government Printing Office.

————, ed. 1881. *Observations of the Transit of Venus, December 8–9, 1874, Made and Reduced under the Direction of the Commission Created by Congress*. Part 2. (Unpublished page proofs held by U.S. Naval Observatory Library.) Washington, D.C.: Government Printing Office.

————. 1897. "Aspects of American Astronomy." *Science* [New Series] 6 (150):709–21.

————. 1903. *The Reminiscences of an Astronomer*. New York: Harper and Brothers.

Newton, Joseph Fort. 1909. *David Swing, Poet-Preacher*. Chicago: Unity Publishing, Abraham Lincoln Centre.

Nichols, Edward L. 1915. "Biographical Memoir of Henry Morton, 1836–1902." In *Biographical Memoirs*, vol. 8, pp. 143–51. Washington, D.C.: National Academy of Sciences.

Noble, Thomas F. X., ed. and trans. 2009. *Charlemagne and Louis the Pious: The Lives by Einhard, Notker, Ermoldus, Thegan, and the Astronomer*. University Park: Pennsylvania State Univ. Press.

Norberg, Arthur L. 1978. "Simon Newcomb's Early Astronomical Career." *Isis* 69 (2):209–25.

Nordgren, Tyler. 2016. *Sun Moon Earth: The History of Solar Eclipses from Omens of Doom to Einstein and Exoplanets*. New York: Basic Books.

Norris, Mary Harriott. 1915. *The Golden Age of Vassar*. Poughkeepsie: Vassar College.

Numbers, Ronald L., and Jonathan M. Butler, eds. 1987. *The Disappointed: Millerism and Millenarianism in the Nineteenth Century*. Bloomington: Indiana Univ. Press.

Osterbrock, Donald E. 1984. *James E. Keeler: Pioneer American Astrophysicist and the Early Development of American Astrophysics*. Cambridge, U.K.: Cambridge Univ. Press.

Owen, Robert Dale. 1849. *Hints on Public Architecture, Containing, among Other Illustrations, Views and Plans of the Smithsonian Institution*. New York: George P. Putnam.

Painter, Nell Irvin. 1977. *Exodusters: Black Migration to Kansas after Reconstruction*. New York: Knopf.

Pang, Alex Soojung-Kim. 2002. *Empire and the Sun: Victorian Solar Eclipse Expeditions*. Stanford: Stanford Univ. Press.

Pasachoff, Jay M. 1999. "Halley and His Maps of the Total Eclipses of 1715 and 1724." *Astronomy and Geophysics* 40 (2):2.18–2.21.

————. 2009. "Solar Eclipses as an Astrophysical Laboratory." *Nature* 459 (7248):789–95.

Peebles, Curtis. 2000. *Asteroids: A History*. Washington, D.C.: Smithsonian Institution Press.

Penrose, F. C. 1878*a*. "The Eclipse of the Sun, 1878, July 29." *Observatory* 2 (18):187–91.

————. 1878*b*. "The Total Eclipse of the Sun, July 29, 1878." *Monthly Notices of the Royal Astronomical Society* 39 (1):48–51.

Peters, C. H. F. 1879. "Some Critical Remarks on So-Called Intra-Mercurial Planet Observations." *Astronomische Nachrichten* 94 (2253):321–36 and (2254):338–39.

Pettit, Michael. 2006. "'The Joy in Believing': The Cardiff Giant, Commercial Deceptions, and Styles of Observation in Gilded Age America." *Isis* 97 (4):659–77.

Phelan, Michael. 1858. *The Game of Billiards*. 3rd ed. New York: D. Appleton.

Pickering, Mary. 1993. *Auguste Comte: An Intellectual Biography*. Vol. 1. Cambridge, U.K.: Cambridge Univ. Press.

Pickering, W. H. 1878. "Total Eclipse of the Sun, July 29, 1878." *Monthly Notices of the Royal Astronomical Society* 39 (2):137–39.

Pleasonton, Gen. Augustus J. 1876. *The Influence of the Blue Ray of the Sunlight and of the Blue Colour of the Sky.* . . . Philadelphia: Claxton, Remsen and Haffelfinger.

Plotkin, Howard. 1977. "Henry Draper, the Discovery of Oxygen in the Sun, and the Dilemma of Interpreting the Solar Spectrum." *Journal for the History of Astronomy* 8 (1):44–51.

———. 1978. "Edward C. Pickering, the Henry Draper Memorial, and the Beginnings of Astrophysics in America." *Annals of Science* 35 (4):365–77.

Proctor, Richard A. 1879. "New Planets near the Sun." *Contemporary Review* 34 (March):660–77.

Raines, Rebecca Robbins. 1996. *Getting the Message Through: A Branch History of the U.S. Army Signal Corps.* Washington, D.C.: Center of Military History.

Randel, William Peirce. 1970. "Huxley in America." *Proceedings of the American Philosophical Society* 114 (2):73–99.

Ranyard, Arthur Cowper. 1878. "Remarks on the Observations of the Eclipse of July 29th, 1878." *Astronomical Register* 16 (190):241–47.

———, ed. 1879. *Memoirs of the Royal Astronomical Society.* Special Issue [Observations Made during Total Solar Eclipses]. 41:1–792.

———. 1881. "Observations of the Total Solar Eclipse of 1878, July 29th, Made at Cherry Creek Camp, near Denver, Colorado." *Memoirs of the Royal Astronomical Society* 46:213–39.

Reed, Helen Leah. 1892. "Women's Work at the Harvard Observatory." *New England Magazine* [New Series] 6 (2):165–76.

Reingold, Nathan. 1963. "A Good Place to Study Astronomy." *Library of Congress Quarterly Journal of Current Acquisitions* 20 (4):211–17.

———. 1964. "Cleveland Abbe at Pulkowa: Theory and Practice in the Nineteenth Century Physical Sciences." *Archives Internationales d'Histoire des Sciences* 17 (67):133–47.

———. 1972. "American Indifference to Basic Research: A Reappraisal." In *Nineteenth-Century American Science*, edited by George H. Daniels, 38–62. Evanston: Northwestern Univ. Press.

Richter, Amy G. 2005. *Home on the Rails: Women, the Railroad, and the Rise of Public Domesticity.* Chapel Hill: Univ. of North Carolina Press.

Rideing, William H. 1876. "At the Exhibition II." *Appletons' Journal* 15 (377):759–60.

Ridenour, Hugh. 2008. "John E. Osborne: A Real 'Character' from the Old West." *Annals of Wyoming* 80 (3):2–16.

Roberts, Philip J. 1981. "Edison, the Electric Light and the Eclipse." *Annals of Wyoming* 53 (1):54–62.

Roscoe, Henry Enfield. 1906. *The Life and Experiences of Sir Henry Enfield Roscoe, D.C.L., LL.D., F.R.S.* London: Macmillan.

Roseveare, N. T. 1982. *Mercury's Perihelion, from Le Verrier to Einstein.* Oxford: Clarendon Press.

Rothenberg, Marc. 1982. *The History of Science and Technology in the United States: A Critical and Selective Bibliography.* New York: Garland.

———. 1986. "History of Astronomy." In *Historical Writing on American Science: Perspectives and Prospects*, edited by Sally Gregory Kohlstedt and Margaret W. Rossiter, 117–31. Baltimore: Johns Hopkins Univ. Press.

Rothschild, Robert Friend. 2009. *Two Brides for Apollo: The Life of Samuel Williams (1743–1817).* Bloomington, Ind.: iUniverse.

Rubin, Vera. 1997. *Bright Galaxies, Dark Matters*. Woodbury, N.Y.: American Institute of Physics.

Rufus, W. Carl. 1938. "A Student Notebook of 1855: The Astronomical Memoranda of the Youthful James C. Watson, Later Professor of Astronomy." *Michigan Alumnus Quarterly Review* 44 (15):141–44.

———. 1951. "The Department of Astronomy." In *The University of Michigan: An Encyclopedic Survey*, edited by Wilfred B. Shaw, vol. 2, pp. 442–65. Ann Arbor: Univ. of Michigan Press.

Ruskin, Steve. 2008. "'Among the Favored Mortals of Earth': The Press, State Pride, and the Eclipse of 1878." *Colorado Heritage* (Spring):22–35.

Ryan, John J. 1879. "Eclipse Expedition of Three Maryland Professors." *Woodstock Letters* 8 (1):25–32.

Sagala, Sandra K. 2008. *Buffalo Bill on Stage*. Albuquerque: Univ. of New Mexico Press.

Sands, B. F. 1870. *Reports on Observations of the Total Eclipse of the Sun, August 7, 1869*. Washington, D.C.: Government Printing Office.

——— 1871. *Reports on Observations of the Total Solar Eclipse of December 22, 1870*. Washington, D.C.: Government Printing Office.

Schmadel, Lutz D. 2003. *Dictionary of Minor Planet Names*. 5th ed. Berlin: Springer.

Schurz, Carl, et al. 1880. *White River Ute Commission Investigation. Letter from the Secretary of the Interior, Transmitting Copy of Evidence Taken before White River Ute Commission, May 14, 1880*. 46th Congress, 2d Session, House of Representatives, Ex. Doc. No. 83. Washington, D.C.: Government Printing Office.

Schuster, Arthur, 1878a. "Some Remarks on the Total Solar Eclipse of July 29, 1878." *Monthly Notices of the Royal Astronomical Society* 39 (1):44–47.

———. 1878b. "The Sun's Corona during the Eclipse of 1878." *Observatory* 2 (20):262–66.

———. 1932. *Biographical Fragments*. London: Macmillan.

Scudder, Samuel Hubbard. 1886. *The Winnipeg Country; or, Roughing It with an Eclipse Party*. Boston: Cupples, Upham and Co.

Shearer, Frederick E., ed. 1882–83. *The Pacific Tourist: An Illustrated Guide to Pacific RR California and Pleasure Resorts Across the Continent*. New York: J. R. Bowman.

Sheehan, William. 1999. "Christian Heinrich Friedrich Peters, September 19, 1813–July 18, 1890." In *Biographical Memoirs*, vol. 76, pp. 288–312. Washington, D.C.: National Academy Press.

Silbernagel, Robert. 2011. *Troubled Trails: The Meeker Affair and the Expulsion of the Utes from Colorado*. Salt Lake City: Univ. of Utah Press.

Simonds, William Adams. 1934. *Edison: His Life, His Work, His Genius*. Indianapolis: Bobbs-Merrill.

Sitton, Thad, and James H. Conrad. 2005. *Freedom Colonies: Independent Black Texans in the Time of Jim Crow*. Austin: Univ. of Texas Press.

Sizer, Nelson. 1882. *The Royal Road to Wealth: How to Find and Follow It. Illustrated by More Than One Hundred Portraits of Those Who Have Achieved Success, Wealth, and Power*. New York: John R. Anderson and Henry S. Allen.

Smallwood, James M. 1981. *Time of Hope, Time of Despair: Black Texans during Reconstruction*. Port Washington, N.Y.: National University Publications / Kennikat Press.

Smith, Phyllis. 1993. *Weather Pioneers: The Signal Corps Station at Pikes Peak*. Athens, Ohio: Swallow Press / Ohio Univ. Press.

Smyth, C. Piazzi. 1853. "On the Total Solar Eclipse of 1851." *Transactions of the Royal Society of Edinburgh* 20 (part 3):503–11.

Sobel, Dava. 2016. *The Glass Universe: How the Ladies of the Harvard Observatory Took the Measure of the Stars*. New York: Viking.

Spear, Robert J. 2005. *The Great Gypsy Moth War: The History of the First Campaign in Massachusetts to Eradicate the Gypsy Moth, 1890–1901*. Amherst: Univ. of Massachusetts Press.

Speiden, Norman R. 1947. "Thomas A. Edison: Sketch of Activities, 1874–1881." *Science* [New Series] 105 (2719):137–41.

Sprague, Marshall. 1957. *Massacre: The Tragedy at White River*. Boston: Little, Brown.

Spring, Agnes Wright. 1968. "Did Edison Get 'Turned On' in Wyoming?" *True West* 15 (5):22–23, 44.

Stamm, Richard E. 2012. *The Castle: An Illustrated History of the Smithsonian Building*. 2nd ed. Washington, D.C.: Smithsonian Books.

Stanley, Matthew. 2012. "Predicting the Past: Ancient Eclipses and Airy, Newcomb, and Huxley on the Authority of Science." *Isis* 103 (2):254–77.

Stanton, Elizabeth Cady, Susan B. Anthony, and Matilda Joslyn Gage, eds. 1887. *History of Woman Suffrage*. Vol. 3: *1876–1885*. Rochester, N.Y.: Susan B. Anthony.

Steel, Duncan. 2001. *Eclipse: The Celestial Phenomenon that Changed the Course of History*. Washington, D.C.: Joseph Henry Press.

Stephens, T. C. 1944. "The Makers of Ornithology in Northwestern Iowa." *Iowa Bird Life* 14 (2):18–37.

Stephenson, F. Richard. 1997. *Historical Eclipses and Earth's Rotation*. Cambridge, U.K.: Cambridge University Press.

Strahorn, Robert E. 1878. *To the Rockies and Beyond, or A Summer on the Union Pacific Railway and Branches*. Omaha: Omaha Republican Print.

Stross, Randall E. 2007. *The Wizard of Menlo Park: How Thomas Alva Edison Invented the Modern World*. New York: Crown.

Swackhamer, Barry A. 1994. "J. B. Silvis, the Union Pacific's Nomadic Photographer." *Journal of the West* 33 (2):52–61.

Sweet, Homer DeLois. 1894. *The Averys of Groton. Genealogical and Biographical*. Syracuse: Rice-Taylor Printing Co.

Swift, Lewis. 1878a. "Discovery of Vulcan." *Science Observer* 2 (2):9–10.

———. 1878b. "The Great Eclipse of 1878." *Science Observer* 1 (9):53–54.

Swing, David. 1889. *Motives of Life*. New and enlarged ed. Chicago: A. C. McClurg.

Talbot, Marion, and Lois Kimball Mathews Rosenberry. 1931. *The History of the American Association of University Women, 1881–1931*. Boston: Houghton Mifflin.

Tate, Alfred O. 1938. *Edison's Open Door: The Life Story of Thomas A. Edison, a Great Individualist*. New York: E. P. Dutton.

Temkin, Owsei, and Janet Koudelka. 1950. "Simon Newcomb and the Location of President Garfield's Bullet." *Bulletin of the History of Medicine* 24 (4):393–97.

Tenn, Joseph S. 1986. "The Hugginses, the Drapers, and the Rise of Astrophysics." *Griffith Observer* 50 (10):2–15.

Thomas, M. Carey. 1908. "Present Tendencies in Women's College and University Education." *Publications of the Association of Collegiate Alumnae* 3 (17):45–62.

Thompson, R. W., and John Rodgers. 1878. *Letter from the Secretary of the Navy, Transmitting, in Answer to a Senate Resolution of February 15, 1878, Information in Relation to the Usefulness of Government Observatories. February 18, 1878*. 45th Congress, 2d Session, Senate, Ex. Doc. No. 28. Washington, D.C.: Government Printing Office

Thorp, Raymond W. 1957. *Spirit Gun of the West: The Story of Doc W. F. Carver*. Glendale, Calif.: Arthur H. Clark.

Tocqueville, Alexis de. 1863. *Democracy in America*. Vol. 2. Translated by Henry Reeve. Edited, with Notes, by Francis Bowen. 2nd ed. Cambridge, Mass.: Sever and Francis.

Todd, Mabel Loomis. 1894. *Total Eclipses of the Sun*. Boston: Roberts Brothers.

Tribble, Scott. 2009. *A Colossal Hoax: The Giant from Cardiff That Fooled America*. Lanham, Md.: Rowman and Littlefield.

Trouvelot, E. L. 1882. *The Trouvelot Astronomical Drawings Manual*. New York: Charles Scribner's Sons.

True, Frederick W., ed. 1913. *A History of the First Half-Century of the National Academy of Sciences, 1863–1913*. Washington, D.C.: National Academy of Sciences.

Twain, Mark, and Charles Dudley Warner. 1874. *The Gilded Age: A Novel*. 3 vols. London: George Routledge and Sons.

Tyndall, John. 1873. *Hours of Exercise in the Alps*. 3rd ed. London: Longmans, Green.

United States Army Signal Corps. 1877. *Annual Report of the Chief Signal-Officer to the Secretary of War for the Year 1877*. Washington, D.C.: Government Printing Office.

———. 1889. *Meteorological Observations Made on the Summit of Pike's Peak, Colorado, Latitude, 38° 50' N., Longitude, 10° 52' W., Height, 14,134 Feet, January, 1874, to June, 1888, under the Direction of the Chief Signal Officer, U.S. Army*. Annals of the Astronomical Observatory of Harvard College, Edward C. Pickering, Director, vol. 22. Cambridge, Mass.: John Wilson and Son.

United States Centennial Commission. 1876. *International Exhibition, 1876, Official Catalogue*. Rev. ed. Philadelphia: John R. Nagle.

United States Naval Observatory. 1878a. *Instructions for Observing the Total Solar Eclipse of July 29, 1878*. Washington, D.C.: Government Printing Office.

———. 1878b. *Instructions for Observing the Transit of Mercury, 1878, May 5–6*. Washington, D.C.: Government Printing Office.

———. 1879. *Reports on Telescopic Observations of the Transit of Mercury, May 5–6, 1878*. Washington, D.C.: Government Printing Office.

———. 1880. *Reports on the Total Solar Eclipses of July 29, 1878, and January 11, 1880*. Washington, D.C.: Government Printing Office.

———. 1885. *Reports of Observations of the Total Eclipse of the Sun, August 7, 1869, Made by Parties Under the General Direction of Professor J. H. C. Coffin, U. S. N., Superintendent of the American Ephemeris and Nautical Almanac*. Washington, D.C.: Government Printing Office.

Upton, Winslow. 1879. "The Solar Eclipse of 1878; a Lecture before the Institute." *Bulletin of the Essex Institute* 11 (4–6):53–71.

Vaeth, J. Gordon. 1966. *Langley: Man of Science and Flight*. New York: Ronald Press.

Van de Warker, Ely. 1872. "Effects of Railroad Travel on the Health of Women." *Georgia Medical Companion: A Monthly Adviser* 2 (4):193–206.

Vassar College. 1881. *The Vassar College Song Book*. New York: Trow's Printing and Bookbinding Co.

Vickers, W. B. 1880. *History of the City of Denver, Arapahoe County, and Colorado*. Chicago: O. L. Baskin.

Vivian, A. Pendarves. 1879. *Wanderings in the Western Land*. London: Sampson Low, Marston, Searle, and Rivington.

Walcott, Charles D. 1912. "Biographical Memoir of Samuel Pierpont Langley 1834–1906." In *Biographical Memoirs*, vol. 7, pp. 245–68. Washington, D.C.: National Academy of Sciences.

Waldo, Leonard, ed. 1879. *Report of the Observations of the Total Solar Eclipse, July 29, 1878, Made at Fort Worth, Texas.* Cambridge, Mass.: John Wilson and Son.

Walker, Francis A., ed. 1880. *United States Centennial Commission. International Exhibition, 1876. Reports and Awards.* Vol. 7: *Groups 21–27.* Washington, D.C.: Government Printing Office.

Warner, Deborah Jean. 1979. *Graceanna Lewis: Scientist and Humanitarian.* Washington, D.C.: Smithsonian Institution Press.

———. 1989. "American Octants and Sextants: The Early Years." *Rittenhouse: Journal of the American Scientific Instrument Enterprise* 3 (11):86–112.

Warner, Deborah Jean, and Robert B. Ariail. 1995. *Alvan Clark & Sons: Artists in Optics.* 2nd ed. Richmond: Willmann-Bell.

Watson, James C. 1877. *American Watches: An Extract from the Report on Horology at the International Exhibition at Philadelphia, 1876.* New York: Robbins and Appleton.

———. 1878*a*. "Discovery of an Intra-Mercurial Planet." *American Journal of Science and Arts* [Third Series] 16 (93):230–33.

———. 1878*b*. "On the Intra Mercurial Planets, from Letters to the Editors, Dated Ann Arbor, Sept. 3d, 5th and 17th, 1878." *American Journal of Science and Arts* [Third Series] 16 (94):310–13.

———. 1879. "Schreiben des Herrn Prof. Watson an den Herausgeber." *Astronomische Nachrichten* 95 (2263):101–6.

———. 1880. "Report of the Director of the Observatory and Professor of Astronomy to the Board of Regents." In *Annual Report of the Board of Regents of the University of Wisconsin for the Fiscal Year Ending September 30, 1880,* 35–36. Madison: David Atwood, State Printer.

Wead, Charles Kasson, et al. 1910. *Simon Newcomb: Memorial Addresses.* Washington, D.C.: Philosophical Society of Washington.

Weber, Timothy P. 1987. *Living in the Shadow of the Second Coming: American Premillennialism, 1875–1982.* Rev. ed. Chicago: Univ. of Chicago Press.

Weeks, Edwin R. 1878. "Personal Observations of the Eclipse." *Western Review of Science and Industry* 2 (5):259–63.

Welch, Walter L. 1972. *Charles Batchelor: Edison's Chief Partner.* Syracuse: Syracuse Univ.

Wellman, Judith. 2004. *The Road to Seneca Falls: Elizabeth Cady Stanton and the First Woman's Rights Convention.* Urbana: Univ. of Illinois Press.

Welsh, Joe, Bill Howes, and Kevin J. Holland. 2010. *The Cars of Pullman.* Minneapolis: Voyageur.

Werrett, Simon. 2010. "The Astronomical Capital of the World: Pulkovo Observatory in the Russia of Tsar Nicholas I." In *The Heavens on Earth: Observatories and Astronomy in Nineteenth-Century Science and Culture,* edited by David Aubin, Charlotte Bigg, and H. Otto Sibum, 33–57. Durham: Duke Univ. Press.

Westfall, John, and William Sheehan. 2015. *Celestial Shadows: Eclipses, Transits, and Occultations.* New York: Springer.

Whitesell, Patricia S. 1998. *A Creation of His Own: Tappan's Detroit Observatory.* Ann Arbor: Bentley Historical Library, Univ. of Michigan.

———. 2003. "Detroit Observatory: Nineteenth-Century Training Ground for Astronomers." *Journal of Astronomical History and Heritage* 6 (2):69–106.

Whitnah, Donald R. 1961. *A History of the United States Weather Bureau.* Urbana: Univ. of Illinois Press.

Williams, Samuel. 1785. "Observations of a Solar Eclipse, October 27, 1780, Made on the East Side of Long-Island, in Penobscot Bay." *Memoirs of the American Academy of Arts and Sciences* 1:86–102.

Willis, Edmund P., and William H. Hooke. 2006. "Cleveland Abbe and American Meteorology, 1871–1901." *Bulletin of the American Meteorological Society* 87 (3):315–26.

Wilson, Curtis. 1989. "The Newtonian Achievement in Astronomy." In *Planetary Astronomy from the Renaissance to the Rise of Astrophysics. Part A: Tyco Brahe to Newton*, General History of Astronomy 2A, edited by René Taton and Curtis Wilson, 233–74. Cambridge, U.K.: Cambridge Univ. Press.

Wilson, Joseph M. 1880. *The Great Exhibitions of the World: with One Hundred and Nineteen Illustrations of Grounds, Plans, Exteriors, &c.* New York: A. W. Lovering.

Wlasuk, Peter T. 1996. "'So Much for Fame!': The Story of Lewis Swift." *Quarterly Journal of the Royal Astronomical Society* 37 (4):683–707.

Woodson, Thomas, ed. 1980. *The Centenary Edition of the Works of Nathaniel Hawthorne.* Vol. 14: *The French and Italian Notebooks.* Columbus: Ohio State Univ. Press.

Wortman, Roy T. 1965. "Denver's Anti-Chinese Riot, 1880." *Colorado Magazine* 42 (4):275–91.

Wright, Carroll D. 1885. *Health Statistics of Female College Graduates.* Boston: Wright and Potter.

Wright, Helen. 1949. *Sweeper in the Sky: The Life of Maria Mitchell, First Woman Astronomer in America.* New York: Macmillan.

Wright, Helena E. 1992. "George Pullman and the Allen Paper Car Wheel." *Technology and Culture* 33 (4):757–68.

Young, Charles A. 1878a. "Observations upon the Solar Eclipse of July 29, 1878, by the Princeton Eclipse Expedition." *American Journal of Science and Arts* [Third Series] 16 (94):279–90.

———. 1878b. "The Recent Solar Eclipse." *Princeton Review* 2 (Nov.):865–88.

Zirker, Jack B. 1995. *Total Eclipses of the Sun.* Princeton: Princeton Univ. Press.

Zschoche, Sue. 1989. "Dr. Clarke Revisited: Science, True Womanhood, and Female Collegiate Education." *History of Education Quarterly* 29 (4):545–69.

ACKNOWLEDGMENTS

F EW REMARKS HAVE SO PROFOUNDLY CHANGED THE TRAJEC- tory of my life. At the time, in 1994, I was working as an NPR science correspondent in Boston, and I had been assigned to report on an annular solar eclipse (for a definition, see page 37) that was to cross nearby New Hampshire. The tenth of May found me in the path of annularity, where, through protective Mylar glasses, I watched the sun metamorphose into a brilliant golden ring. It was an impressive sight, yet Williams College astronomer Jay Pasachoff, a veteran of such conjunctions of the sun and moon, helped put the event in context. In an interview, he explained that a partial solar eclipse, even an annular one, is nothing compared with nature's most awesome show, a *total* solar eclipse. He then offered some advice. "Before you die," he said, "you owe it to yourself, at least once, to experience totality." It was Jay's urging that led me, four years later, to Aruba, where I first encountered the solar corona. My opening acknowledg- ment, therefore, must go to Jay Pasachoff. If not for his remark all those years ago, my passion for eclipses likely would never have been ignited, and without that passion, I surely would never have written this book.

Todd Shuster, my gifted and tireless agent, offered encourage- ment, ideas, and profound patience when I told him, after that initial experience in Aruba, that I aimed to write a book about eclipses— but not just yet. I suggested that we hold off for nineteen years, for what better moment to publish than in the summer of 2017, when the moon's shadow would traverse the United States from the Pacific to the Atlantic for the first time in almost a century. In 2011, Todd and I dusted off my vague book idea, and I began to hunt for a concrete story to tell. I soon discovered the little-known tale of the total solar eclipse of July 29, 1878. Its rich amalgamation of lively characters, weighty themes, and adventurous settings seemed to offer fertile raw material, so I shared this embryonic book concept with my editor, Bob Weil. How I got so lucky to work with such a brilliant man, I will never know. Bob—as usual—saw how to elevate the story far beyond my original notion, motivated me to do my best work, and, when I ultimately handed in the manuscript, took what I wrote and invariably made it better. I cannot adequately express my thanks to Bob, who nurtured this book and remained its unwavering champion.

Pulling off a book that involved so much archival research required a tremendous amount of support, financial and otherwise. I am grateful to the Alfred P. Sloan Foundation, specifically Doron Weber, for generously underwriting this endeavor. My thanks go, as well, to the Charles Redd Center for Western Studies at Brigham Young University, which provided supplemental funding to visit archives in Wyoming and Colorado. Uncountable librarians, archivists, and curators also assisted me in innumerable ways, and while I cannot thank them all by name, some deserve special recognition.

In Washington, Mark Mollan doggedly tracked down items hidden deep inside the National Archives, Richard Stamm led me through the Smithsonian Castle to the site of Joseph Henry's office (where Edison famously exhibited his phonograph in April 1878), and Yann Henrotte gave me a personal tour of the Arts Club of Washington, formerly the home of Cleveland Abbe. Thanks also to Janice Goldblum at the National Academy of Sciences, Gregory Shelton at the U.S. Naval Observatory Library, Norma Rosado-Blake at the American Association for the Advancement of Science, and Tad Bennicoff, Ellen Alers, and Courtney Bellizzi at the Smithsonian Institution Archives.

In Pennsylvania, Marianne Kasica at the University of Pittsburgh helped me sift through Samuel Langley's records from the Allegheny Observatory, and Nancy Miller and Tim Horning at the University of Pennsylvania facilitated my exploration of George Barker's papers. Across the ocean at the University of Exeter, Angela Mandrioli offered similar assistance with the correspondence of Norman Lockyer. Nan Card at the Rutherford B. Hayes Presidential Library and Museums in Fremont, Ohio, and Jascin N. Leonardo Finger at the Nantucket Maria Mitchell Association graciously answered myriad questions about items in their collections. At Vassar College, where many of Maria Mitchell's writings have been archived, Dean Rogers kindly photocopied and forwarded hundreds of pages of relevant material, and Debra Elmegreen and Colton Johnson responded to my queries about the school's illustrious history. On my visit to Poughkeepsie, Fred Chromey showed me the observatory where Maria Mitchell hosted her dome parties, and when I ventured farther upstate to Hamilton College, archivist Katherine Collett opened the papers of C. H. F. Peters to me.

At the University of Michigan, which houses the James Craig Watson Papers, my thanks go to Malgosia Myc and her team at the Bentley Historical Library, and to Karen Wight, who gave me a tour of Watson's old workplace, the Detroit Observatory. At Eastern Michigan University, in nearby Ypsilanti, Norbert Vance permitted me to see—and, thrillingly, to touch—the very telescope that Watson took to Wyoming to find Vulcan, and at The Henry Ford, in Dearborn, Stephanie Lucas gave me a rare behind-the-scenes look at Edison's reconstructed Menlo Park laboratory and an up-close view of an original tasimeter. In Texas, Jessica Baber and Christy Morton of the Layland Museum of History aided me in reconstructing what life was like in Johnson County at the time of the 1878 eclipse. In Iowa, at the Sioux City Public Library, Kim Walish spent hours on my behalf scrolling through streaked and blurry microfilm to find, and then decipher, D. H. Talbot's eclipse report from Wyoming Territory.

In modern-day Wyoming, Larry Brown at the State Archives tracked down many items for me, including the 1881 coroner's inquest into the lynching of "Big Nose George" Parrott. In Rawlins, Palma Jack at the public library shared her deep knowledge of local history, and Corinne Gordon and Carol Reed at the Carbon County Museum allowed me to peruse that institution's archives. Dan Kinnaman, the museum's historian emeritus and author of several books on the region's early settlement, opened his vast personal collection of books and photographs to me. Also connecting me to Wyoming's past was a North Carolinian, Craig Galbraith, who shared items that his great-great-grandfather Robert M. Galbraith had handed down through the generations, including recollections of 1878 Rawlins and of Thomas Edison's visit to town.

As an independent writer unaffiliated with a university, I am blessed to live in a community with a first-rate public library system. At the Boulder Public Library, Ann Berry and Laurel Seppala-Etra cheerfully filled my constant requests for interlibrary loan material, and the team at the Carnegie Branch Library for Local History (Wendy Hall, Hope Arculin, Marti Anderson) handled the many reels of microfilm that arrived from near and far. At the Denver Public Library, Lisa Flavin and Coi Drummond-Gehrig helped me find and scan many of the illustrations in this book. Others who assisted my quest for artwork were Iren Snavely at the State Library of Pennsylvania, Ann Passmore at Penn State, Jennifer Claybourne at the University of Minnesota, Jean Lythgoe at the Rockford Public Library, Maria McEachern at the John G. Wolbach Library of the Harvard-Smithsonian Center for Astrophysics, and Leonard DeGraaf of the Thomas Edison National Historical Park.

As my manuscript neared completion, several experts magnanimously read what I had written to offer advice, critiques, and corrections. Paul Israel, director of the Thomas A. Edison Papers at Rutgers, helped ensure that passages describing the life of the great inventor remained true to the historical record. Daniel Kevles, historian of science at Yale, and Deborah Jean Warner, history of science curator at the Smithsonian National Museum of American History, assessed my characterizations of late nineteenth-century American scientists and scientific institutions. Mark Miller, former Wyoming state archaeologist and author of *Hollow Victory: The White River Expedition of 1879 and the Battle of Milk Creek*, vetted what I wrote about frontier Rawlins and the Meeker Massacre. Dr. Peter Hackett, director of the Institute for Altitude Medicine in Telluride, Colorado, reviewed my medical assessment of Cleveland Abbe's ailing condition in the rarefied air atop Pikes Peak. And astronomer Jay Pasachoff, already mentioned for his role in sparking the creation of this book, scrutinized the final product for scientific accuracy. He and his wife, the author Naomi Pasachoff, meticulously read my manuscript and offered many suggestions for improvement. I cannot adequately express my gratitude to this panel of authoritative readers. Any errors that have slipped through despite their advice are fully my own.

I am fortunate to be surrounded by good friends who are also great writers and editors. Early on, when I first conceived this book, conversations with Dan Glick, Tim Weston, Andrea Meyer, and Dana Meyer valuably shaped my thinking. When I then fashioned these ideas into a book proposal, Kathryn Bowers and Andy Bowers offered key insights that helped me frame the story for a broad audience. During the long months of research and writing, Jonny Waldman drew me out of my solitary existence for happy-hour literary discussions that helped sustain my enthusiasm and focus my thoughts. When I finally had a draft manuscript to share, Len Ackland, Alison Richards, Katy Human, and my father (Charles Baron) read it carefully and offered pivotal advice for adjusting the narrative where it veered off course. Rhitu Chatterjee, Carol Stutzman, and Ken Bader read later iterations and helped fine-tune the manuscript. I am also immensely grateful to copyeditor Annie Gottlieb, who, with good humor and scrupulousness, cleaned up my lapses of language and fixed my logical inconsistencies. Thanks, as well, to Marie Pantojan at Liveright, who deftly guided my book through editing and production and held my hand through each step of the complex process.

Friends and family assisted me in many practical ways. Dave Groobert genially hosted me on my research trips to Washington. Gloria Cohen, in New Haven, visited Yale University's Manuscripts and Archives on my behalf. John Miller generously donated his full set of the *Encyclopaedia Britannica* from the 1870s, which assisted my research. I received foreign-language help from Deborah Keyek-Franssen, who creatively translated the poem "One Little Vulcan" from the German, and Alison Perlo, who assisted my reading of the French writings of Auguste Comte. Alan and Mary Frankel accompanied me on an eclipse chase in Munich in 1999. Rachel Nowak was also with me that year and then, in 2012, joined me for totality in Queensland along with her delightful daughters, Ting and Jie Giovannitti-Nowak, whose presence made that eclipse especially joyful.

Others who have sustained me with their encouragement and moral support include Dianne; Jessica; Ira, Sharon, Sophia, and Max; Sam and Aubrey; Jason, Cheryl, Leah, Cassie, and Jason Jr.; Glenn and Sharon; Amy and Philip; Jane; Spider and Louise; Ruthy and Karen; Mary-Alice and Walter; Ailsa and Kate; John and Carole; Richard and Valerie; Bill and Michael; Gilles and Emma; Robin and Grant. Above all, I am grateful to the person who has been by my side on this entire adventure, from the beach in Aruba through this book's publication: my husband, Paul Myers, who served as my primary reader, confidant, sounding board, and cheerleader. He shared my excitement when things went well, calmed me in the occasional moment of panic, and enriched the journey—and life in general—in so many ways.

Two people who deserve my deep gratitude I unfortunately never had the opportunity to meet. One is the late Jack Eddy, a noted solar astronomer who was himself fascinated by the total solar eclipse of 1878. His published writings (see bibliography) and unpublished lecture notes (held by the archives of the National Center for

Atmospheric Research, in Boulder) guided me as I traveled back in time. I similarly benefited from the scholarship of Patricia "Sandy" Whitesell, a historian at the University of Michigan who had been working on a biography of James Craig Watson when she died, far too young. Her husband, John Wolfe, kindly shared her unpublished manuscript and research notes, which proved invaluable. I so wish I could have met Sandy and Jack. We would have had much to discuss.

I never more palpably felt the presence of individuals long gone than in late July 2015, when I ventured into the Wyoming desert west of Rawlins in search of Separation, the railroad stop long since abandoned. Local historian Dan Kinnaman and his wife, Angie, were my guides, and we were joined by Jack Eddy's daughter Amy Gale (and her daughter Cami), who with her family in 1968 had excavated two stone-and-concrete piers built by the scientific team at Simon Newcomb's eclipse camp. It took us several hours of searching in the heat and thorny brush, but we eventually located the piers, now largely buried by the shifting sand. For Amy, those stone markers brought memories of her father. For me, they were monuments to the remarkable people who had stood at that spot and elsewhere along the path of totality on July 29, 1878. I am grateful to Maria Mitchell, James Craig Watson, Thomas Edison, Simon Newcomb, Norman Lockyer, Cleveland Abbe, and so many others for putting their thoughts to paper. I hope this book does justice to their lives.

INDEX

Page numbers in *italics* refer to illustrations.

ABOUT THE AUTHOR

DAVID BARON, recent chair in astrobiology at the Library of Congress, worked in public radio for almost thirty years as a science correspondent for NPR, science reporter for Boston's WBUR, and science editor for the program *The World*. His journalism has earned honors from the Overseas Press Club of America, the National Academy of Sciences, and the American Association for the Advancement of Science. His previous book, *The Beast in the Garden*, received the Colorado Book Award. He lives in Boulder, Colorado.